Silkworm Egg Science

Principles and Protocols

Silkworm Egg Science

Principles and Protocols

– by–

Dr. Tribhuvan Singh
Science-C
Research Extension Centre
Central Silk Board, Ministry of Textiles
Govt. of India, Rampur Road
Una (Himachal Pradesh)

Madan Mohan Bhat
Scientist-D
Regional Sericultural Research Station
Central Silk Board, Ministry of Textiles
Govt. of India, Sahaspur
Dehradun (Uttarakhand)

Dr. Mohammad Ashraf Khan
Director
Central Sericultural Research & Training Institute
Central Silk Board, Ministry of Textiles
Govt. of India, Pampore
(Jammu & Kashmir)

2018

Daya Publishing House®

A Division of

Astral International Pvt. Ltd.

New Delhi - 110 002

First Published, 2010
Reprinted, 2018

ISBN: 978-81-7035-663-9 (Hardbound)

Publisher's note:

Every possible effort has been made to ensure that the information contained in this book is accurate at the time of going to press, and the publisher and author cannot accept responsibility for any errors or omissions, however caused. No responsibility for loss or damage occasioned to any person acting, or refraining from action, as a result of the material in this publication can be accepted by the editor, the publisher or the author. The Publisher is not associated with any product or vendor mentioned in the book. The contents of this work are intended to further general scientific research, understanding and discussion only. Readers should consult with a specialist where appropriate.

Every effort has been made to trace the owners of copyright material used in this book, if any. The author and the publisher will be grateful for any omission brought to their notice for acknowledgement in the future editions of the book.

Published by : **Daya Publishing House**®
A Division of
Astral International Pvt. Ltd.
– ISO 9001:2008 Certified Company –
4736/23, Ansari Road, Darya Ganj
New Delhi-110 002
Ph. 011-43549197, 23278134
E-mail: info@astralint.com
Website: www.astralint.com

Preface

Silk enters the 21st century as strong, smooth, much-sought-after environment friendly, healthy textile fiber. Despite shrinking of sericulture in Japan and Korea due to high cost industrialization and severe competition from manmade, silk has not only survived but has succeeded globally in raising its production and demand. Indian sericulture today represents a picture of an established, flourishing and well organized industry spread all over the country, made a phenomenal progress in silk production and emerged as the second largest producer of silk in the world only after China. There is growing demand for quality silk in both domestic and international markets, for which quality seed (egg) is the vital input. Silkworm eggs the 'anchor sheet of silk industry' plays a decisive role in the productivity, production, quality and development of silk industry.

'What we sow is what we harvest', literally means that the quality of seed we use, determine the quality and yield of harvest. The region, season and technology adopted also decide the quality and quantity of harvest. Crop success, thus, is decided by a number of factors including the quality of seed and silkworm crop is not an exception. How to define silkworm seed quality? It is disease freeness to pathologists; performance true to the original breed characters as per expected vigour to a breeders; high yield of good quality cocoons to a farmer; trouble free reeling, expected silk recovery and high grade quality silk to a reeler and so on. Seed quality against the sericulture industrial background is, therefore, the fulfillment of all these expectations in a broader way.

If the quality of egg is good, one can harvest a successful crop and if it is inferior, the crop could be lost at any stage of the silkworm rearing depending on the magnitude of the defects and or inadequacies of the seed. Quality is the product of technology and inputs. Input refers to inter alia breed, quality of seed cocoons, survival rate, pupation rate, disease freeness, maintenance of optimum temperature, humidity, ventilation and light, infrastructure facilities, technical manpower availability, technology adopted etc. Quality seed is also to be true to its breed with better recovery, rich in fecundity with high fertility rate, uniform hatching and above all ensures successful stable cocoon crop.

Production of silkworm seed is a commercial activity and how much seed the country needs of course depends on how efficient the rest of sericulture is. Naturally, if per hectare productivity or per 100 Dfls productivity of cocoons is better, the country can do with less seed because after all, the limiting factor is the mulberry acreage which cannot be increased horizontally forever and under such circumstances, we will have to think of vertical expansion. However, given a particular quantity of leaf or acreage under mulberry, it is not possible to increase productivity by using more seed. Requirement of more seed per unit area in India is because of low productivity per 100 Dfls. The average productivity/100 Dfls in China are approximately 70.00 kg but the productivity in India is only 51.78 kg/100 Dfls. This indicates that the country is consuming more layings compared to China to produce the same quantity of cocoons. If we had higher productivity, we could reduce the egg production. In other words, we are taking probably more of a burden than required by producing more layings and making farmers to consume more seed per unit area to produce cocoons at the lower yield level. However, if we are able to reduce the gap to the reasonable degree of fairness, India can reduce the gaping hiatus in the production levels of the two global competitors.

Among the various aspects of sericulture science, the one on 'silkworm eggs' deals with the information required for planning the egg production and preservation of eggs on scientific lines by adopting technological advancements achieved on the subject. Production of silkworm seed, which is not only free from all kind of diseases but also ensure the purity and vigour of the breeds to exploit heterosis, involves various technologies at various levels of egg production. Further, adoption of techniques for egg production on a mass scale, to meet the huge demand of silkworm seed in the country, differs from those of small scale in the laboratories.

Therefore, it was thought desirable to publish a comprehensive book on the subject. The present publication "*Silkworm Egg Science : Principles & Protocols*" is designed as a self-contained document. In this book, the theoretical as well as practical aspects of silkworm eggs have been explained lucidly. It is designed to address wide range of readers, experts, university teachers, students, researchers, technologists, policy makers and to fulfill the hopes and aspirations of all those engaged in sericulture in general and egg production in particular. It will serve as a very useful guide and will motivate and assist in the efforts to find solutions to the existing field problems to increase production, productivity and quality of silk.

The book has many unique features. It comprises chapters on disinfection and hygiene, egg production technologies, embryonic development, egg diapause and biochemical modulations, preservation of eggs, hibernation schedules, artificial hatching of eggs for rearing at appropriate time and season, methods of incubation, adaptation to changing environment besides prevention and control of pebrine disease of eggs and pests to achieve sustained and quality silkworm eggs. The technical components is meticulously incorporated in this book to make it fairly comprehensive reference material which will serve in a long-way to improve the quality of seed in the days-to-come at all levels of silkworm egg production. Our efforts will bear fruits, if the status of silkworm egg production in India is raised to a level which will be at par with number one silk producing country in the world.

To present the comprehensive and up-to-date information in this book "*Silkworm Egg Science: Principles & Protocols*" we have drawn necessary information and scores from the published books, documents, periodicals, journals, technical bulletins, reports etc. We gratefully acknowledge with thanks this valuable debt. We profoundly and sincerely are deeply privileged and honored by Ms M. Sathiyavathi, IAS, Member Secretary, Central Silk Board, Bangalore for encouragement and blessings to write this book. We deeply express our gratitude to Dr. Rajat Mohan, Scientist-C, Regional

Sericultural Research Station, Sahaspur, Dehradun, Dr. Ranjan Tiwari and Dr. S. K. Tyagi, Scientist-C, Research Extension Centre, Una, and Sri N. M. Biram Saheb, Scientist-D, Silkworm Seed Production Centre, Central Silk Board, Mysore whose co-operation and involvement enabled us in bringing out the book in this form. We strongly hope that this book will serve as a guideline for all those involved in both upstream and downstream processes of egg production.

It is with great pleasure, we wish to express our sincere thanks to all those we had interactions and enlightened discussions which stimulated and prompted us to take up this onerous task of compiling all the available information on the 'Silkworm Eggs'. We wish to place on records our profound thanks to all those friends and well wishers who have directly or indirectly helped in one way or the other and for the constant encouragement in bringing out this publication.

We profoundly thank Sri Anil Mittal, Publisher, Daya Publishing House, New Delhi whose interest and enthusiasm has brought out this book with an excellent presentation within a short period.

Though all out efforts have been made to incorporate correct and authentic information, but it may not be out of place to mention that some errors might have crept in, which is not intentional but may be due to oversight and constraints. Please bring out such errors to the notice of authors and publishers. Views and suggestions from users of this book are welcome to improve upon future editions.

Tribhuwan Singh

Madan Mohan Bhat

Mohammad Ashraf Khan

Contents

Chapter 1

INTRODUCTION

Sericulture in India is viable agro based industry aptly matches with the socio-economic back-drop of the rural region of the country. Silk, the queen of fibres, is the fruitful culmination of the cumulative and concerted efforts of the multi-disciplinary functionaries like cocoon producers, grainueres (egg producers) and reelers, weavers etc. India is credited for at least four distinctions in the world of silk. Indians are the largest consumers of silk, second largest producer of silk, largest importer of mulberry raw silk and producers of all four commercially exploited silk in the world *viz.*,–Mulberry, Tasar, Eri and Muga and has been recording consistent growth in the quality, production and productivity. As India encompasses wide geographical and agro-climatic variations, mulberry sericulture is distributed in temperate, sub-tropical and tropical zones, while the major share comes from the tropics. Prior to independence, mulberry raw silk production was only 800 MT per year and sericulture was considered as a subsidiary occupation of the poor farmers. In fact, popularization of improved agronomical practices, release of improved mulberry varieties, evolution of productive silkworm breeds and hybrids coupled with crop protection measures and appropriate rearing and reeling technologies changes the whole sericulture scenario in 1980. Total shift from the use of pure multivoltine silkworm eggs to MV x BV or BV x BV eggs and replacement of local indigenous mulberry varieties by improved varieties took place within short period of time. The contribution of R&D resulted increase in cocoon production/ha from 294 kg/ha/year in 1975–76 to about 702 kg/ha/yr in 2008–09. Raw silk productivity/ha increased from 20.34 kg/ha in 1975–76 to 87.72 kg/ha in 2008–09. The renditta, which is the most important parameter to judge the quality of silk, has come down from 14.5 kg in 1975–76 to 8.0 kg in 2008–09. The raw silk production has now reached to 15,610 MT from 2,541 MT in 1975–76 (Table 1.1). The credit for this achievement goes equally to the present day sericulturists with a receptive mind, willing to readily take up the challenges for better quality and productivity on one hand and to the pioneering efforts of Indian sericulture scientists and extension personnel on the other, who are striving hard to develop suitable region and season specific technologies, and take them to the field. The success is rather the cumulative result of different factors like better viability of sericulture over other cash crops, introduction of high yielding mulberry varieties and silkworm breeds, improved extension support system and increased acceptance of various package of practices by sericulturists. Today, Indian sericulture represents a picture of an established, flourishing and well organized industry spread all over the country, made a phenomenal progress in silk production and emerged as the second largest producer of silk in the world only after China.

These have been incessant endeavors to improve the silk both in terms of quality and quantity. One of the inputs that play a decisive role in the success of silkworm crop is the silkworm egg. If the quality of egg is good, one can harvest a successful crop and if it is inferior, the crop could be lost at any stage of the silkworm rearing depending on the magnitude of the defects and or inadequacies of the seed. Quality is the product of technology and inputs. Input refers to silkworm breeds, quality of seed cocoons, survival rate, pupation rate, disease freeness, maintenance of optimum temperature, humidity, aeration and light, infrastructure facilities, technical manpower availability, technology used etc. Quality seed is also to be true to its breed with better recovery, rich in fecundity and high fertility rate, uniform hatching and above all would ensures successful stable cocoon crop.

Silkworm seed is the anchor sheet of silk industry and the production of silkworm seed is a commercial activity. One of the

Table 1.1: Mulberry Sericulture in India

Year	Area Under Mulberry (ha)	Dfls Production (Lakh Nos)	Reeling Cocoon Production (MT)	Raw Silk Production (MT)	Renditta	Raw Silk Productivity/ ha (kg)	Cocoon Productivity/ ha (kg)	Cocoon Production/ 100 Dfls (kg)
1975–76	124913	1706.90	36739	2541	14.5	20.34	294.12	21.52
1980–81	170000	2189.51	58208	4593	12.7	27.02	342.40	26.58
1985–86	217839	2628.79	76717	7029	10.9	32.27	352.17	29.18
1990–91	316610	3171.88	116663	11486	10.2	36.28	368.48	36.78
1991–92	331237	3183.43	1071.53	10658	10.1	32.18	323.49	33.66
1992–93	342764	3402.44	129685	13000	10.0	37.93	378.35	38.12
1993–94	319215	3034.32	117268	12550	9.3	39.32	367.36	38.65
1994–95	283093	3189.70	123115	13450	9.2	47.51	434.89	38.60
1995–96	286496	2856.20	116362	12884	9.0	44.97	406.16	40.74
1996–97	280651	3199.49	115655	12954	8.9	46.16	412.10	36.15
1997–98	282244	3355.20	127495	14048	9.1	49.77	451.72	38.00
1998–99	270069	3318.62	126565	14260	8.9	52.80	468.64	38.14
1999–00	227151	3309.80	124531	13944	8.9	61.39	548.23	37.62
2000–01	215921	3182.86	124663	14432	8.6	66.84	577.35	39.17
2001–02	232076	3337.00	139616	15842	8.8	68.26	601.60	41.84

Contd...

Table 1.1–Contd...

Year	*Area Under Mulberry (ha)*	*Dfls Production (Lakh Nos)*	*Reeling Cocoon Production (MT)*	*Raw Silk Production (MT)*	*Renditta*	*Raw Silk Productivity/ ha (kg)*	*Cocoon Productivity/ ha (kg)*	*Cocoon Production/ 100 Dfls (kg)*
2002–03	194463	2920.10	128181	14617	8.8	75.17	659.15	43.90
2003–04	185120	2598.10	117471	13970	8.4	75.46	634.57	45.21
2004–05	171959	2565.40	120027	14620	8.2	85.02	698.00	46.79
2005–06	179065	2625.50	126261	15445	8.2	86.25	705.11	48.09
2006–07	191893	2838.10	135462	16525	8.2	86.12	705.92	47.70
2007–08	184928	2647.50	132038	16245	8.1	87.84	714.00	49.87
2008–09	177943	2411.00	124838	15610	8.0	87.72	701.56	51.78

MT: Metric Tonnes.

initial and important steps in the silkworm rearing is to produce and supply examined disease free layings (Dfls) to the rearers. The task of evolving basic seed and maintaining parent stock and the operation of further multiplication and distribution requires a highly scientific nucleus technical skill manpower and efficient organization and therefore a rational approach is required in silkworm seed production. Silkworm eggs are produced either for reproduction of seed (parental seed) or for industrial seed (commercial seed). The parental seeds are maintained by Central and State Government Departments. Requirement of commercial seed in the country of-course depends on the efficiency of rest of sericulture activity. Therefore, naturally, if per hectare cocoon productivity and or per 100 Dfls productivity of cocoons is better in the country, the requirement can be met with less seed. Because after all, the limiting factor is the production of leaf or the mulberry acreage. However, given a particular quantity of leaf or acreage under mulberry, it is not possible to increase productivity by using more seed. It is very well established and understood that consuming of more seed per unit area is due to low productivity per 100 Dfls. The average productivity/100 Dfls in China are approximately 70.00 kg in comparison to 51.78 kg/100 Dfls in India. This clearly reveals that the country is consuming more layings compared to China per unit area to produce the same quantity of cocoons. However, if we could achieve higher productivity, we may be able to reduce the egg production. In other words, probably we are taking more of a burden than required by producing more layings and making farmers to consume more layings per unit area to produce cocoons at the lower yield level.

In sericulturally advanced countries, the seed production technologies have been perfected to its highest level. Though, in India silkworm seed production is practiced since decades, it is yet to be standardized to its perfection and efforts are being made in this direction. To meet the requirement of Indian sericulture farmers to produce 15,610 MT of raw silk or 1, 24,838 MT of reeling cocoons, the country at present consuming about 24 crore disease free commercial silkworm seed. Karnataka is the major silk producing state in the country, which needs about 10 crores of disease free layings (Dfls) per year. The major production of seed (about 74 per cent) comes from Licensed Seed Producers (LSPs) while Centre contributes about 10 per cent of the total seed requirement in the country and remaining 16 per cent is contributed from State sectors *i.e.* Directorate of Sericulture (DOS) of respective States. Timely supply of required quantity of silkworm seed to the sericulturists is one of the prerequisite for achieving stable cocoon crop, higher productivity and production. Scientific/systematic seed multiplication and generation of genetically pure silkworm seed cocoons for preparation of commercial silkworm seed could achieve this. Despite technological advancements and transfer of technologies to the field, the desired level of improvement is not still achieved in the field. The main reasons are inadequacies associated with the extension activities, poor infrastructure and socioeconomic conditions of the sericulturists. As a result, the seed cocoons produced by the seed farmers are not in conformity with the characters of pure breeds. Besides, due to climatic vagaries, coupled with field problems, the generation of seed cocoon is more often not commiserating with the requirements, resulting in diversion of large quantum of seed cocoons to the reeling sector and non-availability of required quantity of seed during the rearing season.

The factors responsible for deterioration of quality of seed cocoons are poor quality of leaf, unhygienic conditions during rearing, non-maintenance of optimum micro-climatic conditions (temperature, humidity, photoperiod, ventilation) and lack of knowledge besides low-level technology adoption. To overcome all these problems, efforts are being focused for the improvement of seed quality by:

- ✰ Adoption of better management strategies for seed cocoon procurement.
- ✰ Adoption of effective and efficient seed cocoon transportation technology.
- ✰ Adoption of sex separation technology at cocoon stage to ensure desired hybrid seed production and or introduction of sex limited breeds at seed cocoon production level.
- ✰ Adoption of low temperature preservation technology for seed cocoons, pupae and moths if required for synchronization during mating, oviposition and egg production.
- ✰ Adoption of standardize efficient technologies, techniques and practices of silkworm seed production.

- ☆ Popularization of loose egg production technology and to ensure its demand and supply to commercial silkworm rearers.
- ☆ Adoption of appropriate egg preservation schedules as per requirements.
- ☆ Introduction of eco-friendly custom designed equipments for grainages.
- ☆ Introduction of high productive breeds to ensure better recovery and profit.
- ☆ Adoption of exclusive package of practices of mulberry cultivation and silkworm rearing for seed cocoon generation.
- ☆ Adoptions of price fixation strategy for seed cocoons based on pupation rate/live pupa percentage.
- ☆ Quarantine of silkworm seed/laying for disease freeness.
- ☆ ISO certification of silkworm seed producing centers for quality maintenance.

Seed production organized on scientific lines commenced with the establishment of Central Silk Board (CSB), a statutory Government organization to coordinate sericulture activities and silk industry in the country. The CSB has setup a National level institute for research on seed 'Silkworm Seed Technology Laboratory' (SSTL) at Bangalore to address various problems encountered in seed production and evolve efficient technologies of seed production and National Silkworm Seed Organization (NSSO) at Bangalore. Out of 1229 Silkworm Seed Production Centers (SSPCs), 182 are under state sector, 23 under NSSO (central sector) and 1024 under Licensed Seed Producers (LSPs) (private sector). Central Silk Board has also established 14 Basic Seed Farms (BSFs) for producing and distributing parental breeds of silkworm for further multiplication and distribution as per requirement throughout the country. The seed production scenario in Indian sericulture presents dichotomy. Seed organization can be categorized into reproductive seed maintenance and management and its multiplication up to quality hybrid seed production and its distribution for heterosis exploitation. Under Indian conditions, the organization of seed production coupled with region and season specific development has to be addressed to meet domestic requirements and quality orientation.

Table 1.2: Silkworm Seed Production in Different States in India

(Unit: Lakh numbers)

Year	*Karnataka*	*Andhra Pradesh*	*Tamil Nadu*	*West Bengal*	*J&K*	*Others*	*Total*
1980–81	1603.4	89.0	64.1	390.0	27.7	15.2	2189.5
1985–86	1703.9	158.2	146.1	676.0	42.6	40.8	2767.7
1990–91	2112.1	249.2	195.0	490.0	35.8	89.7	3171.9
1991–92	1912.4	378.8	162.3	598.5	41.5	89.9	3183.4
1992–93	2012.6	416.6	213.6	621.7	37.7	100.2	3402.4
1993–94	1962.1	236.3	152.6	572.6	19.4	91.3	3034.3
1994–95	2178.0	154.7	130.7	621.8	26.2	78.2	3189.7
1995–96	1768.2	206.9	212.6	532.6	33.2	102.7	2856.1
1996–97	1845.0	516.5	175.4	548.2	36.9	77.5	3199.5
1997–98	1886.3	627.4	139.7	600.2	31.6	69.9	3355.2
1998–99	1859.6	722.3	139.5	487.7	35.6	73.9	3318.6
1999–00	1726.4	810.8	153.0	507.2	36.2	76.3	3309.8
2000–01	1602.2	854.0	153.0	458.2	33.5	81.9	3182.9
2001–02	1690.0	879.0	131.5	529.0	33.0	74.5	3337.0
2002–03	1287.1	970.5	82.3	485.8	30.1	64.3	2920.1

Contd...

Table 1.2–Contd...

Year	*Karnataka*	*Andhra Pradesh*	*Tamil Nadu*	*West Bengal*	*J&K*	*Others*	*Total*
2003–04	974.4	1016.6	40.4	489.1	28.1	49.5	2598.1
2004–05	1094.1	848.1	55.7	486.1	30.9	50.5	2565.4
2005–06	1088.9	882.9	88.9	480.7	32.4	51.7	2625.5
2006–07	1221.9	901.4	130.5	492.0	24.7	67.6	2838.1
2007–08	1183.9	711.6	154.6	483.1	28.7	85.6	2647.5
2008–09	943.0	723.0	153.0	461.3	29.8	100.9	2411.0

Maintenance and multiplication of breeders' stocks is organized into tiers with or without selection pressures at different levels keeping primarily in view not to allow variations in qualitative as well as quantitative characters of the parental breeds, hence for true to mother multiplication up to mass production of industrial seed for determined goals.

In India, a systematic three-tier seed multiplication system is lacking in most of the states. In Karnataka, which is the major silk and silkworm seed producing state in the country has well established three-tier system of seed organization for seed multiplication and distribution. Other states may replicate the Karnataka model for systematic seed production and supply. Mostly, the supply of silkworm seed in India is taken by the following agencies:

- ☆ Grainages run by the State Department of Sericulture under state sector.
- ☆ Grainages run by the Central Silk Board under central sector.
- ☆ Other institutes recognized by Government *viz.* Karnataka State Sericulture Research & Development Institute (KSSRDI), Asian Institute of Rural Development (AIRD) etc.
- ☆ Seed Producers who are licensed (Licensed Seed Producers) (LSPs) under private sector for the production and supply of commercial hybrid silkworm seed.

Of the total silkworm seed produced in the country (2411.00 lakh Dfls), about 39.11 per cent is produced in the state of Karnataka (943.00 lakh Dfls) (Table 1.2) and 29.98 per cent in Andhra Pradesh (723.00 lakh Dfls). The cocoon productivity which was 21.52 kg/ 100 Dfls during 1975-76 have shown increasing trend and now reached to 51.78 kg (2008-09) reveal the fact of adoption of new technologies of both silkworm rearing and mulberry cultivation, great awareness among farmers, adoption of hygienic conditions, incubation procedure, seed production and preservation technologies, prevention and control of diseases and pests etc.

Chapter 2

DISINFECTION AND HYGIENE

The silkworm is affected by a large number of diseases caused by virus, bacteria, fungi and protozoa. These diseases are known to occur in almost all Sericultural regions of the world. The adverse environmental factors like temperature, humidity and poor quality of mulberry leaves reduces the tolerance of the host to the pathogens and hence increases susceptibility to infections. Disinfections and hygiene forms an integral part of healthy and successful silkworm rearing and egg production. Crop losses due to incidence of diseases are one of the major problems encountered by sericulturists in India. The incidence of silkworm diseases and crop losses is greatly influenced by rearing practices, frequency of cropping, conditions of rearing and grainage houses, general hygiene and environmental conditions favcurable for pathogen build up and spread. Many pathogens, especially fungal and bacterial spores are light and can be drifted easily by air current, leading to spread of the disease. They have also the ability to remain alive in the environments for longer periods.

Cocoons arriving in the grainages from different sericulture regions of the country can form a continuous source of pathogen entry into seed production units. Processing of batches continuously without providing sufficient time gap, leads to inadequate disinfection and thereby favours pathogen build up in the grainages. Some of the facts about silkworm diseases are:

- All the infectious diseases of silkworms are caused by pathogens.
- The adverse environmental factors like temperature, humidity and feeding poor quality of mulberry leaf reduces the tolerance level of host to the pathogens and hence increases susceptibility to infections.
- The diseases of silkworms are highly contagious and spread very fast.
- There is no silkworm breed/hybrid, which is completely resistant to all the diseases.
- The silkworm pathogens survive longer in the rearing and grainage environment and remain infective.
- The diseases have atypical and typical morphological symptoms.

The disinfection can be defined as destruction of disease causing germs. It is the total destruction of diseases caused by pathogens. There is no curative method for any of the silkworm diseases and hence they can best be prevented rather than cured. This can be achieved by adoption of proper and effective methods of disinfection before and after every silkworm crop and grainage operations. Therefore, to prevent crop losses due to diseases, disinfection is inevitable. Avoiding the spread of silkworm diseases and production of disease free silkworm seed in the grainage is as much important as checking the spread of diseases in silkworm rearing. Continuous production of seed without sufficient gap between the two batches does not provide opportunity for regular disinfection and therefore resulting in buildup of pathogens in the grainages. Silkworm seed is the basic material determining success or failure of a cocoon crop and therefore adopting proper disinfection and hygienic method in preparation of silkworm seed is the foremost step towards raising a healthy silkworm crop.

Since, the species of microorganisms vary and situation in which they may occur differ greatly, no one or two methods are generally applicable. Each situation is a problem in itself and the methods employed depend on the knowledge, ingenuity and purposes of the operator. There are four main reasons for killing, removing or inhibiting microorganisms to maintain hygienic conditions. They are:

- ☆ To prevent infection to man, his animals and plants.
- ☆ To prevent spoilage of food and other commodities.
- ☆ To prevent interference by contaminating microorganisms in various industrial processes that depends on pure culture.
- ☆ To prevent contamination of materials used in pure culture work in laboratories (diagnostics, research, industry etc.).

Common methods of killing or removing microorganisms are:

- ☆ Destruction by heat (boiler, oven etc), chemical agents (disinfectants), radiation (X-ray, ultraviolet rays etc.), mechanical agents (crushing, shattering by ultrasonic vibration) etc.
- ☆ Removal (especially bacteria) by filtration, high seed centrifugation etc.
- ☆ Inhibition by low temperatures (refrigeration, dry ice), desiccation (drying process), high osmotic pressures (lymph, brines etc), chemicals and drugs etc.

Selection of Disinfectants

Several disinfectants are available in the market but only few are widely used. The disinfectants available in the market can be grouped as:

- ☆ Halogens (Chlorine, Iodine etc.).
- ☆ Heavy metals ($HgCl_2$, Mercurochrome, Metaphen, Protargol, etc.).
- ☆ Phenol compounds (Lysol, Creosols etc.).
- ☆ Alcohols (Ethyl and Isopropyl).
- ☆ Formaldehydes (strong reducing agent which inactivates even enzymes).
- ☆ Ethylene oxide (Carboxide, Cryocide etc.).

Qualities of an Ideal Disinfectant

Ideal disinfectant is selected taking into consideration of various factors. The ideal disinfectant must posses the following qualities:

- ☆ Highly effective against a wide variety of micro-organisms in concentration as low as to be economical for use as well as harmless to human beings, domestic animals and non-toxic to plants.

- ✰ Non-injurious and non-staining to materials like fabrics, furniture or metal wares and non-offensive to odour or taste.
- ✰ As specific as possible for micro-organisms.
- ✰ Does not damage the building, materials, appliances and equipments.
- ✰ A good surface tension reducer (have good wetting and penetrating properties).
- ✰ Stable in storage.
- ✰ Readily available in market and less expensive.
- ✰ Easily applied under household or other practical conditions of use.
- ✰ Completely microbicidal within a few minutes or an hour at the most and not inducing macrobiotics, leading to a false sense of security.
- ✰ Non-corrosive.

But, no single disinfectants have all these ideal properties. Some agents may be ideal under some conditions but not under others, *e.g.*, Creosol may be ideal for floors or sanitary purposes but harmful to infants.

Conditions Necessary for Effective Disinfections

It is established that disinfectants respond better under some specific conditions whereby disinfection becomes perfect and effective, such as:

Hydration

Considerable quantity of water is required to facilitate the action of disinfectants, *e.g.*, dehydrated protein to coagulate is difficult since it will turn brown or char. Moreover, resistance of bacterial endospore to heat is probably caused in part by their extremely dehydrated conditions.

Time

No disinfectant, as ordinarily used, acts instantly. Sufficient time for contact must be allowed for whatever chemical and physical reactions to occur. The time required will depend on the nature of the disinfectant, concentration, pH and temperature, nature of target organism and existence of the bacterial population of cells having varying susceptibilities to the disinfectant.

Temperature

With respect to the microbicidal action of heat, temperature is inversely related to time. In case of chemical microbicides, as a rule, the warmer the disinfectant, the more effective is its action. Higher temperatures generally reduce the surface tension, increase acidity, decrease viscosity and diminish adsorption.

Concentration

Effectiveness of a disinfectant is generally related to concentration exponentially, but not linearly. For example, doubling a 0.5 per cent concentration of phenol in aqueous solution does not merely double the killing rate. Doubling the concentration again may increase the effect by only a negligible amount. There is clearly an optimum concentration of phenol at about 1 per cent . Thus, a concentration of a disinfectant beyond a certain point accomplishes increasingly less and is wasteful.

pH

An important factor in detergency is pH which is a measure of acidity or alkalinity. As the pH of the solution increases *i.e.* becomes more alkaline, the efficiency of the product increases. There are limits to this, as too high pH will frequently result in deleterious effects on the surface being cleaned or disinfected. In general, a pH not higher than 10.5 is recommended. As a general rule, the lethal or toxic action of harmful agents involving physical and chemical actions is increased by increased concentration of 'H' or 'OH'–ions.

Osmotic Pressure

Fluids of high osmotic pressure (*e.g.* food preserving syrups and brines) tend to dehydrate the cell contents and so increase resistance of microbial cells to heat and chemical disinfectants.

Surface Tension

Surface tension is of basic importance in disinfection. There are two aspects of this factor, adsorption of surface disinfectants or interfering substances on the surface of the cells and the effect of disinfectants on the wetting and spreading properties of the solution. Both affect contact between disinfectants and microorganisms.

Materials Required for Disinfections

Disinfectants, detergent, sprayer, buckets, measuring jars, weighing scale (balance), gas masks, slaked lime, hand gloves and muslin cloths are required during disinfection.

Method of Disinfections

Disinfection methods are classified broadly of two types; physical and chemical:

A] Physical Methods (Burning, burying and exposure to sunlight)

The physical method includes the use of physical agents such as ultraviolet radiation, flame, dry heat and streams etc. In sericulture, exposing the grainage equipments to bright sunlight for 8–10 hrs is a very economical method of disinfection under Indian conditions. Heat is the major destroyer. Dry heat destroys the microorganism by oxidation and is non-corrosive. Dry heat requires longer exposure time and or higher temperature. Use of boiling water is another effective method. The maximum temperature of 100°C can be obtained and exposure of 10–30 minutes is found effective. Moreover, it is cumbersome and somewhat non-reliable method. However, best results can be achieved by a combination of both physical and chemical methods.

B] Chemical Methods

Chemical means of disinfection involves only those chemicals which have germicidal activity against targeted microbes. The use of chemical for disinfection achieved momentum since 1867, when Lister stated that phenol (carbolic acid) would kill micro-organism. From then onwards, various chemicals have been tested for their germicidal activities and few of them are found effective in sericulture also.

Formalin

It is the most commonly used disinfectants in all sericultural countries. Formalin is aqueous solution of formaldehyde. It is commercially available in the form of 36–40 per cent formaldehyde. The specific gravity is 1.081 in 36 per cent and 1.087 in 40 per cent formalin. The molecular formula of formalin is CH_2O. The disinfecting action of formalin is due to reducing process of formaldehyde *i.e.* HCOH + O = HCOOH. Formaldehyde takes oxygen out of the germ cells leading to death of the germs. A mixture of 2 per cent formalin + 0.5 per cent slaked lime is very effective solution that can be used for disinfection purpose. Formaldehyde upon storage polymerizes into trioxymetylene, which is ineffective as spray and therefore storage of formaldehyde for long time is not advisable. The action of formalin takes place under wet conditions and therefore, the surface of equipments and walls should be drenched with the solution. The action of formalin is faster and more pronounced at temperature above 25°C and humidity > 70 per cent . The action is greatly reduced at temperature below 20°C. This mixture is more effective only if grainage houses/rooms/buildings could be closed to near airtight conditions. In India, more than 90 per cent of the rearing and grainage buildings are not suitable for disinfection with formalin as they are open type dwelling houses. However, formalin is also reported to be a carcinogen and therefore use of it as far as possible should be avoided

Bleaching Powder

Bleaching powder is also called chlorinated lime. It is white amorphous powder with a characteristic of pungent odour of chlorine. The efficacy of bleaching powder is very much dependent on the level of active chlorine in the compound. For effective disinfection, a high-grade bleaching powder with active chlorine content of 30 per cent and above must be used. Bleaching powder when comes in contact with water produces weak acid (HCl) and releases nascent oxygen which has strong oxidizing action against germ cells. It should be stored in sealed bags, away from moisture and light to avoid degradation which rendered it ineffective. Freshly prepared solution of bleaching powder gives the best results. A 2 per cent bleaching powder in 0.3 per cent slaked lime is used for disinfection as spray. The action of bleaching powder is optimal under wet conditions and therefore, surface of appliances, equipments, walls of grainage rooms should be drenched with this solution.

Slaked Lime (Calcium Hydroxide)

It is very widely used bed disinfectant and drying agent in sericulture. It absorbs moisture and can be used to regulate bed humidity and maintain hygiene in grainage buildings. It has strong antiviral action. Application of lime dust alone or in combination with bleaching powder in and around grainage houses and premises improves hygiene in the environment. It is prepared from burnt lime stone or shells by sprinkling water followed by pulverizing and sieving. Lime stone or $CaCO_3$ is burnt to produce quick lime (CaO) which when hydrated forms slaked lime (Calcium hydroxide) [$Ca(OH)_2$] which is used as disinfectant in sericulture.

Paraformaldehyde

It is white crystalline substance with a strong odour of formalin formed by polymerization of formaldehyde. It is effective for fumigation purpose and is an ingredient of certain bed disinfectants. On heating, it sublimates and releases formaldehyde gas which inactivates the pathogens in wet condition and therefore the humidity of the room should be simultaneously increased to maintain the optimum level of requirement for grainage and rearing.

Chlorine Dioxide

Chlorine dioxide (ClO_2) is marketed in different names such as sanitech, serichlor etc. It is an ideal disinfectant suitable for all types of grainage buildings. In combination with slaked lime, it is effective against all silkworm pathogens. It is a strong oxidizing agent, 2.5 times stronger than chlorine and two times stronger than sodium hypo-chloride. It is effective at broader ranges of pH and less reactive with organic compounds. It is least corrosive and non-hazardous. Chlorine dioxide (Sanitech) 500 ppm in 0.5 per cent slaked lime may be used for disinfection. It is stable and may be activated at the time of its use. It possesses tolerable odor and least corrosive at recommended concentration.

Quantity Requirement of Disinfectants

Quantity requirement of disinfectant is estimated based on the surface area of the rearing/grainage building or trays to be disinfected. The quantity requirement may be estimated as:

(*a*) Room: (Total area in square meter)

Area of roof and floor: L x B x 2

Area of two opposite walls: L x B x 2

Area of two other side walls: L x B x 2

(*b*) Grainage trays (wooden/plastic): L x B x 2 x Number of trays

(*c*) Bamboo trays: $\Pi r^2 \times 2 \times$ Number of trays

The disinfectant solution should be prepared @ one liter per 2.5 square meters.

$$\text{Solution required in liters} = \frac{\text{Total area in square meters}}{2.5}$$

Preparation of Disinfectant Solutions

Disinfectant solution should be prepared a fresh before use.

1. 2 per cent Formalin Solution

Commercial formalin contains 36 per cent formaldehyde. This is diluted to prepare 2 per cent solution of formalin as follows:

$$\frac{\text{Concentration of available formalin} - \text{Concentration of formalin required}}{\text{Concentration of required formalin}} = \text{Parts of water to be added in one part of formalin}$$

$$i.e. \quad \frac{36-2}{2} = 17 \text{ parts of water}$$

$$\text{or, } N_1V_1 = N_2V_2 \text{ or } V_1 = \frac{N_2V_2}{N_1} \quad i.e.\ V_1 = \frac{2\text{x}100\ (\text{ml})}{36} = 5.55$$

100.00 – 5.55 = 94.45 ml water; Ratio = 94.45 : 5.55 = 17 : 1

where,

N_1: Given strength of solution

N_2: Desired strength of solution

V_1: Volume of stock chemical required

V_2: Desired volume of solution to be prepared

Practically, for all disinfection purposes, one part of formalin can be mixed with 17 parts of water to prepare 2 per cent formalin solution (Table 2.1) from commercially available formalin (36 per cent).

Table 2.1: Preparation of 2 per cent Formalin Solution of Different Quantities from 36 per cent Stock Solution of Formalin

Sl.No.	*Total Volume of Solution Required (Liter)*	*Quantity of Formalin Required*	*Water to be Added (Liter)*	*Ratio of Formalin Water*
1.	1.00	56 ml	0.944	1: 17
2.	5.00	280 ml	4.720	1: 17
3.	10.00	560 ml	9.440	1: 17
4.	15.00	840 ml	14.160	1: 17
5.	20.00	1.120 liter	18.880	1: 17
6.	25.00	1.400 liter	23.600	1: 17
7.	30.00	1.680 liter	28.320	1: 17
8.	35.00	1.960 liter	33.040	1: 17
9.	50.00	2.800 liter	47.200	1: 17
10.	75.00	4.200 liter	70.800	1: 17
11.	100.00	5.600 liter	94.400	1: 17

If commercial formalin contains 36 per cent formaldehyde, ratio of formalin and water in preparation of solution of different concentration of formalin will be as follows (Table 2.2).

2. Formalin 2 per cent + 0.5 per cent Slaked Lime Mixture

After preparation of 2 per cent formalin solution, slaked lime is added to it @ 5 gm in one liter of formalin solution. Slaked lime is prepared by sprinkling water on burnt limestone and pulverizes it into fine powder of 200–250 mesh size.

Estimation of Quantity of Disinfectant Required

The quantity required for disinfection may be calculated adopting the following procedure (example):

Length of floor (L) = 20'

Breadth of floor (B) = 15'

Floor area for disinfection = L x B = 20' x 15' = 300 square feet or 28 m^2.

Table 2.2: Preparation of Different Concentrations of Formalin

Sl.No.	*Concentration of Formalin Required (Per cent)*	*Concentration of Available Formalin (Per cent)*	*Ratio of Water: Formalin*
1.	1	36	35.00 1
2.	2	36	17.00: 1
3.	3	36	11.00: 1
4.	4	36	8.00: 1
5.	5	36	6.20: 1
6.	6	36	5.00: 1
7.	7	36	4.10: 1
8.	8	36	3.50: 1
9.	9	36	3.00: 1
10.	10	36	2.60: 1

The disinfectant required for disinfection of rearing house is @ 2 liter/square meter floor area or 185 ml/sq. feet floor area. Therefore, disinfection solution required = Total area x @ 2 liter/sq. meter or 28 m^2 x 2 = 56 liter

Disinfection for outside the rearing house and appliances is to be added in this quantity. Additional disinfectant required for appliances is estimated to be 25 per cent of the solution required for floor area *i.e.* 56 x 25 per cent = 14 liters.

Additional disinfectant required for outside the rearing/ grainage house is estimated to be 10 per cent of the solution required for floor area *i.e.* 56 x 10 per cent = 5.6 liters. Hence, total quantity of disinfectant required = 56 + 14 + 5.6 = 75.6 or 75 liters (Table 2.1).

3. 5 per cent Bleaching powder solution (Stable 30 per cent chlorine)

Solution of 5 per cent bleaching powder is prepared by dissolving 50 gm of the chemical in one liter of water (ratio; 1: 20).

The mixture is agitated well. It is filtered through a layer of muslin cloth and the clear solution is used for spraying/disinfection. A ready reckonor for preparation of 5 per cent bleaching powder solution of different volumes is presented in Table 2.3.

Table 2.3: Preparation of Different Quantities of 5 per cent Bleaching Powder Solution

Sl.No.	Total Volume of Solution Required (Liter)	Bleaching Powder Required (gm)	Water to be Added (Liter)	Ratio of Bleaching Powder : Water
1.	1.00	50	1.00	1: 20
2.	5.00	250	5.00	1: 20
3.	10.00	500	10.00	1: 20
4.	15.00	750	15.00	1: 20
5.	20.00	1000	20.00	1: 20
6.	25.00	1250	25.00	1: 20
7.	30.00	1500	30.00	1: 20
8.	35.00	1750	35.00	1: 20
9.	50.00	2500	50.00	1: 20
10.	100.00	5000	100.00	1: 20

The ratio of powder and water to be mixed for preparation of different concentrations of bleaching powder solution is presented in Table 2.4.

$$\text{Quantity of chemical required} = \frac{\text{Per cent of solution to be prepared} \times \text{Quantity of solution required}}{100}$$

Examples

1. 2% of 1 liter solution $= \dfrac{2 \text{ per cent} \times 1000 \text{ (ml)}}{100} = 20$ gm

2. 5% of 1 liter solution $= \dfrac{5 \text{ per cent} \times 1000 \text{ (ml)}}{100}$ 50 gm

3. 2% of 75 liter solution = $\frac{2x\ 75,000\ ml}{100}$ =1500 gm or 1.5 kg

Table 2.4: Requirement of Bleaching Powder for Preparation of Different Concentration of Solution

Sl.No.	*Concentration of Solution Required (Per cent)*	*Quantity of Water (Liter)*	*Quantity of Powder (gm)*	*Ratio of Bleaching Powder: Water*
1.	1.00	1.00	10.00	1: 10
2.	2.00	1.00	20.00	1: 50
3.	3.00	1.00	30.00	1: 33
4.	4.00	1.00	40.00	1:25
5.	5.00	1.00	50.00	1:20
6.	6.00	1.00	60.00	1:17
7	7.00	1.00	70.00	1: 14
8.	8.00	1.00	80.00	1: 13
9.	9.00	1.00	90.00	1: 11
10.	10.00	1.00	100.00	1: 10

4. Bleaching Powder-Slaked Lime Mixture

High-grade bleaching powder is mixed with finely powdered 200–250 mesh size slaked lime @ 50 gm per 950 gm of lime (1:19) to prepare 5 per cent bleaching powder solution. The mixture is used for dusting at the entrance and grainage building surroundings to maintain proper hygiene.

5. Slaked Lime Powder

Burnt rock lime (limestone) or shell lime is procured, water is sprinkled and it is allowed to become powder. This is further pulverized, sieved and made into the form of fine powder, which can be readily used. Application of lime dust in and around grainage building and premises improves hygiene in the environment.

6. Chlorine Dioxide (Sanitech 20,000 ppm stable)

It is powerful, users' friendly, non-hazardous, safe and effective disinfectant for silkworm rearing and grainage house and equipments. Unlike formalin, it is non-carcinogenic. It is capable of successfully destroying microbial and all silkworm pathogens. Sanitech is 2.5 times more potent than chlorine and 50 times more effective than any hypochlorite like bleaching powder. Demands no airtight conditions of rearing and grainages houses for its action. There is no limitation of temperature and humidity for its action. Sanitech do not have corroding action on metallic items and rearing equipments at the recommended concentration. It remains in association with water as oxygen does in water. Sanitech is effective on a broad pH range (pH: 5–9). It does not give pungent smell or suffocation. Chlorine Dioxide (500 ppm) in 0.5 per cent slaked lime can be prepared as:

Solution A

Add 50 gm of activator crystal to 500 ml Sanitech (ClO_2) solution to activate chlorine dioxide, stir and allow 10 minutes for complete dissolution of crystals. Colour changes to light yellow. Add the prepared 500 ml yellow solution to 19 liters of water to get 19.5 liter of solution

Solution B

Dissolve 100 gm of slaked lime powder in 500 ml water in another clean container and allow for some time to settle down.

Mix solution A and B to obtain a total of 20 liters of disinfectants. One liter of Sanitech makes 40 liters of disinfectants. For disinfection of one square meter of area, two liter of Sanitech solution is required. To prepare 2.5 per cent Chlorine Dioxide + 0.5 per cent slaked lime solution of different quantities, the requirement will be as detailed in Table 2.5.

Disinfection of Grainage Houses, Appliances etc.

Drench all parts of grainage building inside and outside and appliances uniformly using gutter or jet sprayer with required quantity of disinfectant (@ 2.0 liter/m^2 floor area of rearing/grainage house + 25 per cent of disinfectant solution for appliances + 10 per cent for outside of the building). After disinfection, grainage houses must be closed for a minimum period of 24 hrs. Cocoon storage trays, oviposition trays should not be smeared with cow dung.

Disinfected grainage houses must be opened at least 24 hrs before arrival of seed cocoons from seed cocoon market for seed production.

Table 2.5: Preparation of 2.5 per cent Chlorine Dioxide + 0.5 per cent slaked lime solution

Sl.NO.	*Volume of Solution Required (Liter)*	*Quantity of Sanitech Required (Liter)*	*Water to be Added (Liter)*	*Activator Crystal Required (gm)*	*Quantity of Lime to be Added (gm)*
1.	5.00	0.125	4.875	12.50	25.00
2.	10.00	0.250	9.750	25.00	50.00
3.	20.00	0.500	19.500	50.00	100.00
4.	30.00	0.750	29.250	75.00	150.00
5.	40.00	1.000	39.000	100.00	200.00
6.	50.00	1.250	48.750	125.00	250.00
7.	60.00	1.500	58.500	150.00	300.00
8.	80.00	2.000	78.000	200.00	400.00
9.	100.00	2.500	97.500	250.00	500.00

Dipping the Appliances in Disinfectant

Disinfect the grainage appliances in 2 per cent bleaching powder in 0.3 per cent slaked lime solution by dipping for at least 10 minutes in disinfection tank. The disinfection tank of 2 feet depth and 4 feet diameter is considered suitable for disinfection purpose. Prepare the disinfectant solution of required quantity and concentration and fill half of the tank. To determine the volume of the tank and the disinfectant solution to be prepared the formula llr^{2}h is used. Example:

The volume of tank of 4 feet diameter and 2 feet height

Diameter = 4 feet and therefore radius (r) = 4 ÷ 2 = 2; as only half of the tank is to be filled, the height of solution in the tank will be 2 ÷ 2 = 1'

Therefore, llr^{2}h = 3.14 x 2 x 2 x 1 = 12.56 cubic feet

One cubic feet holds 28 liters of solution, therefore 12.56 cubic feet will hold

12.56 X 28 = 352 liters of solution.

To prepare 352 liters of 2 per cent bleaching powder in 0.3 per cent slaked lime solution, the requirement will be as follows:

Requirement of water = 352 liters

Requirement of bleaching powder = 352 x 20 gm = 7.040 kg.

Requirement of 0.3 per cent slaked lime = 352 x 3 gm = 1.056 kg.

Prepare 352 liters of solution and fill in the disinfection tank. Dip the appliances as many as possible to submerge in the solution. After 10 minutes the appliances are removed from the disinfection tank and carried directly to the disinfected grainage houses. Afterwards, another set of appliances is dipped in the solution. This system will continue for 8–10 times. The solution is discarded after using 10 times and fresh solution is prepared for the purpose.

Maintenance of hygienic conditions is ensured by rigorous practice of disinfection on the vulnerable foci of infection. Disinfection of grainage room and appliances to prevent vertical transmission of pathogens is very much essential which is to be carried out before grainage operation with available disinfectants to destroy pathogens. Disinfection of other sources of contamination/infection to prevent horizontal transmission of pathogens is:

- ✰ Surface sterilization of silkworm eggs.
- ✰ Disinfection of grainage rooms/building.
- ✰ Disinfection of surrounding areas of grainage building.
- ✰ Disinfection of grainage appliances, oviposition and storage room.
- ✰ Personal hygiene.

Before to use the disinfectants, it is of utmost importance to think about the residual effect of the disinfectants towards environmental pollution and positive side effects on the persons who are handling the disinfectants as also their pets. As no disinfectant is absolutely free from the above, only those disinfectants are to be selected where the side effects are minimal. At present Chlorine dioxide ranks first being the most eco-friendly and users friendly disinfectant available to the sericulture industry. Regular practice of disinfection in all possible ways has proved to be very much effective to maintain hygiene of silkworms and its successful implementation will minimize crop loss due to diseases.

Maintenance of Hygiene during Grainage and Egg Production

- ☆ Do not use grainage appliances without disinfection.
- ☆ Avoid overlapping grainage operations.
- ☆ Seed cocoons should be spread over sheet papers in trays which are disinfected to avoid contamination.
- ☆ Melted and dead cocoons should be separated from a batch and subjected for microscopical examination.
- ☆ Hand gloves, aprons and foot wears should be used while handling seed cocoons, pupae, moths etc. in the grainages.
- ☆ Maintain personnel and grainage hygiene throughout grainage operation.
- ☆ Restrict the entry of persons in the grainage house.
- ☆ Disinfect the silkworm eggs with 2 per cent formalin for 5–10 minutes and wash with clean water, dry them in shade. Disinfect also before initiation of incubation.
- ☆ Moth testing area should be properly disinfected every day before and after carrying out the microscopical examination.
- ☆ At the passage of entrance to grainage building, sprinkle 5 per cent bleaching powder in slaked lime @ 18 gm/sq.ft.
- ☆ Grainures or any other person involved in grainage operation entering to the grainage rooms must disinfect foot and hand before entry. Hand should be thoroughly washed with alkaline germicidal soap.
- ☆ The foot mat place at the entrance of grainage building is to be soaked in 2 per cent bleaching powder in 0.3 per cent slaked lime solution. The foot mat must be soaked in disinfectant solution daily.
- ☆ Separate footwear and clean apron should be used.
- ☆ Wipe the floor daily after grainage with 2 per cent bleaching powder in 0.3 per cent slaked lime solution. If the floor is of mud, dust 5 per cent bleaching powder in slaked lime at an interval of 3–4 days without fail.
- ☆ Moth testing equipments such as crushing sets, mixie cups, glass rods, beakers, funnels etc. should be cleaned and disinfected after every sample.

Table 2.6: Disinfectants and their Requirement for 200 sq. ft. Floor Area of Grainage Building

Item to Disinfectant	*Mode of Disinfection*	*Quantity Required*	*Quantity of Disinfectants Required*		
			Bleaching Powder (kg)	*Slaked Lime (kg)*	*Detergent (kg)*
Grainage house (200 sq. ft. or 18.58 sq. mtr) and appliances (8 oviposition stands and 160 trays)	Spray disinfection, cleaning and washing with 2 per cent bleaching powder in 0.3 per cent slaked lime + 50 per cent for appliances	Floor area of grainage house x 2 liters/sq. mtr + 50 per cent for appliances *i.e.* 18.58 x 2 = 37.00 + 18.50 liters = 55.5 liters	1.11	0.166	–
Disinfection of grainage appliances by soaking the trays, plastic sheets etc	2 per cent bleaching powder in 0.3 per cent slaked lime solution	500 liters.	10.00	1.500	–
Spray disinfection of grainage house and surroundings	0.3 per cent slaked lime solution + 25 per cent for surroundings	37 liters + 9.25 liters = 46.25 liters	–	0.138	–
Spray disinfection of grainage house and appliances	2 per cent formalin + 0.5 per cent detergent + 50 per cent for appliances + 25 per cent for rearing house and surroundings	37 liters + 18.5 liters + 9.25 liters = 64.75 liters or 65.00 liters	–	–	0.325

- ☆ The grainage waste like crushed moths, melted cocoons, sheet papers etc. should be treated with 2 per cent bleaching powder solution and deposited in a disinfection pit (soak pit) which must be away from the grainage building.
- ☆ Cocoons should not be stored nearby the seed processing area and separate stored room should be provided for them.
- ☆ The grainage staff should properly be trained in different procedures of disinfections and maintenance of hygiene.

Chapter 3

PRODUCTION OF EGGS

Production of silkworm seed is an intense and tedious job. Every arrangement is to be made well in advance to avoid any confusion, error or hindrance during preparation of laying. The grainage building must be composed of seed cocoon preservation room, low temperature room and silkworm egg preservation room. The rooms for emergence of moths, mating, egg laying and preservation of silkworm eggs should be equipped with air conditioning devices to regulate temperature, humidity and ventilation as per requirement. The rooms should be uniformly bright during daytime and dark during night. Strong light and wind should be avoided. The perforated cotton cloth/paper for covering the cocoons, the trays for mating of male and female moths and egg laying should be cleaned and disinfected properly before every use. All the materials should be in sufficient number/quantity which, must correspond with the highest rate of daily emergence and laying preparation.

Seed cocoons are the source to produce silkworm eggs. The quantity of seed cocoons to be purchased for production of eggs is related to the quantity of silkworm eggs to be produced besides success of rearing of parent seed. Therefore, seed cocoons harvested have to be subjected to strict visual and microscopical examination. The cocoons that qualify the fixed norms should be procured to ensure production of high quality silkworm eggs, stable cocoon crop and better yield.

Seed Cocoons Transportation

The seed cocoons are harvested on 7th or 8th day of spinning depending on the season, breed and other extrinsic and intrinsic factors. Seed cocoons after harvesting from mountages are subjected for elimination of melted, Uzi infested and defective cocoons at farmer's level itself before transportation to seed cocoon market and from there to egg producing centers (grainages). These cocoons loosely packed in plastic perforated crates/baskets are transported using seed cocoon transportation vehicle to avoid vigorous jerk and shaking of cocoons. Transport of seed cocoons packed in gunny bags are to be discouraged. It is better to transport either early in the morning or late at night when out side temperature is slightly lower. Mid-day, when the temperature is usually high should be avoided for transporting the seed cocoons. Also transportation of seed cocoons to long distances should be avoided. Pack only 8–10 kg of seed cocoons in each plastic basket. This helps in minimizing the physical trauma to pupae, which results in increased melting at later stages.

Purchase of Seed Cocoons

Only after completion of pupation, the seed cocoons should be purchased. Seed cocoons should be free from all kind of diseases in general and from pebrine disease in particular. Thorough inspection of disease freeness is required before purchase of cocoons for seed production. Purchase only those batches which, fulfill the following norms:

- ·Disease freeness.
- Pupation rate above 90 per cent .
- Defective cocoons less than 5 per cent .
- Number of cocoons/kilogram in bivoltine must be in between 570–650 only.
- Cocoons must confirm the breed characters.
- Average yield of cocoons/100 Dfls must be above 45.00 kg.

If the cocoons are not within the norms, have to be rejected and sent for reeling and in no case compromise with inputs, which are essential to produce quality silkworm seed.

Selection of Seed Cocoons

After quality inspection, seed cocoons are subjected for individual selection, and those, which do not confirm to the racial characteristics of the parental breeds, should be discarded. This maintains the uniform quality of seed cocoons to be utilized for egg production. Defective cocoons which are to be rejected are thin shelled cocoons, malformed cocoons, fluffy cocoons, light yellow or yellowish or other coloured cocoons, melted cocoons, calceiform cocoons, stained cocoons, Uzi-infested cocoons, mute cocoons, urinated cocoons etc. All defective cocoons which are unfit for seed preparations are stifled in hot air chamber and sent for reeling. If this is not done, such cocoon gives foul smell and contaminates the grainage. Further, such cocoons attract demestid beetles, which may also eat healthy moths in the grainages during egg production. Defective cocoons are formed mostly due to genetic characteristics or unfavourable environmental factors during incubation and seed crop rearing. Any hybrid cocoons if found in the lot, which is to be utilized for egg production are also to be forthwith rejected.

Preservation of Seed Cocoons

Improper method of preservation of seed cocoons lead to irregular emergence. The conditions required for preservation of seed cocoons are as follows:

Temperature

Seed cocoons are to be preserved at recommended temperature. Optimum temperature for preservation of bivoltine seed cocoon is 25±1°C, at which the emergence rate is high, number of eggs laid by each female moth, is maximum and rate of dead and unfertilized eggs is lower. If seed cocoons are preserved at 30°C, the rate of moth emergence decreases; number of unfertilized eggs, dead eggs and non-hibernating eggs increases significantly besides affecting the preserved pupae with flacherie disease. Male pupae are more affected at high temperature. If temperature is lower than 20°C, moth emergence rate decreases, fecundity fall sharply, number of unfertilized eggs increases, synchronization becomes difficult due to prolongation of moth emergence and here female pupae suffers the most. The effect of different preservation temperatures during pupal stage on egg production is detailed in Table 3.1.

Table 3.1: Relationship Between Different Preservation Temperatures during Pupal Stage and Egg Production

Preservation Temperature (°C)		*No. of Eggs Laid*		*No. of Good Eggs*		*No. of Unfertilized Eggs*	
Before Pupation	*After Pupation*	*Fecundity*	*Index*	*Per Moth*	*Index*	*Per Moth*	*Index*
25	25	576	100	552	100	24	100
20	20	480	83	448	81	32	133
30	30	246	44	172	31	74	308
20	30	394	68	311	56	83	345
30	20	326	58	300	54	26	108

Source: Ming *et al*., 1989.

Humidity

The optimum humidity considered for preservation of seed cocoon is 75–80 per cent . Moreover, it is established that humidity in the range of 65–90 per cent has no evident adverse effect on the health of moth, fecundity and number of unfertilized eggs and on the voltinism of silkworm in the next generation. But when humidity is less than 60 per cent, the pupal development becomes slow, emergence percentage of moth decreases and number of moths failing to mate also increases. Therefore, humidity during preservation of seed cocoons should be adjusted to optimum recommended level.

Photoperiod

Cyclical light and darkness is to be maintained *i.e.* care should be taken to make brightness in the daytime and darkness at night. It is important that the night be comparatively dark and the preserved cocoons are exposed to light before dawn on the day of eclosion/ emergence.

Ventilation

Preservation room should be such that no heavy winds and direct sunlight should fall on it. The rate of respiration of pupae inside the cocoons increases with the development. The rate of respiration and release of carbon dioxide is high when compound eyes of pupae becomes coloured. Air should be kept fresh through ventilation in the preservation room.

Correlation Between Pigmentation and Development of Pupae

Non-synchronization of spinning dates and erratic moth emergence is very common due to varying rearing conditions. Therefore, to regulate synchronization of emergence of moths for mating, it is necessary to observe the morphological changes in growth and development of pupae regularly. Temperature should be regulated in accordance with the degree of pigmentation of the compound eyes, antennae and their body in order to synchronize the emergence of moths of two silkworm varieties simultaneously for hybrid preparation.

Generally, pigmentation and development of pupae are correlated (Ming *et al.*, 1989) as:

- ☆ The start of pigmentation of compound eyes marks the half waypoint between mounting and emergence.
- ☆ The compound eyes become black at the two-third point.
- ☆ When the antennae turn dark black, moth will emerge in two or three days.
- ☆ When the pupae body becomes loose and soft, loses its luster and becomes shriveled, moths will emerge the following day.
- ☆ If non-synchronization is noticed, consign female pupae at 7°C for 1 or 2 days only, a day before moth emergence.

Quality Inspection of Seed Cocoons

Seed cocoons purchased from the market for seed production are subjected for quality inspection as indicated below:

- ☆ Number of cocoons per kilogram.
- ☆ Cut open half kilogram of seed cocoons and record percentage of defective cocoons *viz.*, melted, dead, Uzi infested and muscardine affected.
- ☆ Percentage of double cocoons.
- ☆ Percentage of live pupae.

- ☆ Percentage of male and female cocoons.
- ☆ Assess cocoon weight, shell weight and shell ratio by taking 25 cocoons at random.

Quality assessment of seed cocoons will help in ascertaining the rejection percentage of seed cocoons and also the male and female ratio.

Factors Influencing Seed Quality

Several factors contribute to the ultimate quality of silkworm seed. Quality silkworm seed must be processed from good parental silkworm crop and cocoons confirming to the norms, processed and handled scientifically, free from seed borne pebrine and other diseases and capable of giving successful cocoon crop under standard conditions. The major important factors are:

- ☆ Stable and productive silkworm breeds capable of producing superior quality of raw silk. Maintenance and multiplication of the breeds, true to the breed characters.
- ☆ Parental silkworm seed cocoon generation, which are free from all kind of diseases in general and transovarially transmitted pebrine disease in particular and fit as per norms for seed cocoon preparation.
- ☆ Hybrid silkworm seed production and their handling following scientific methods of testing and procedures.
- ☆ Perfect planning of seed production and utilization.

Arrangement of Cocoons

Good cocoons selected are arranged in single layer either in wooden, plastic or bamboo trays. Bamboo trays of 105 cm diameter can hold 1000–1200 multivoltine or 800–900 bivoltine seed cocoons. Overcrowding of cocoons should be avoided as it leads to pupal mortality. After separating the cocoons according to their sexes, they are arranged separately in preservation room/chambers. For preservation of pupae corrugated sheet or paddy husk is used. In order to increase the efficiency of separating the cocoons and moths perforated paper/cloth is used to cover the tray a day before the expected date of start of emergence. However, if cocoons are cut and pupae are separated from the cocoons, they are preserved in plastic trays (60 cm x 90 cm size) using corrugated sheet at the base. In each tray, 750–800 female pupae and 850–900 male pupae can be accommodated. Cover the cocoons/pupae with moth picking nets or old newspaper with holes of 2.5 cm diameter at a distance of 10 cm. To avoid mixing, preserve different sexes of different breeds in separate rooms. Provide optimum conditions of temperature (25 ± 1°C), RH (80 ± 5 per cent) and photoperiod (16 L: 8 D).

Developmental events that occur once in the life cycle of an insect have long been known to express at specific part of the day to manifest a population rhythm. Emergence can be predicted by recognizing the changes in the colouration and flexibility of the body of the pharate adult, dark pigmentation of the compound eyes, antennae and wings and loss of luster of pupal skin and its elasticity, soft nature of body and formation of wrinkles upon touch.

Sex Separation

Almost all the silkworms laying produced in India for commercial cocoon production are from the hybrids (Single crosses or double crosses). To produce hybrids, it is inevitable to separate male and female sexes of pure breeds before emergence and or copulation. Sex separation is carried out with the aim of:

- ☆ To prepare desired hybrids of silkworm.
- ☆ To synchronize emergence of male and female silkworm moths for mating.
- ☆ Cocoon assessment and basic seed productions.

Methods of Sex Separation

Sexes are separated at various stages of development with different techniques. They are separated at egg stage by sex-limited egg colour, at larval stage by imaginal buds for sexual organs and sex limited larval markings, at cocoon stage by cocoons colour and pupal morphological features, size and weight besides distinct developmental trends in male and female larvae. Seed cocoons reaching in grainages are sex separated primarily by morphological features as highlighted below (Table 3.2).

Cocoon Cutting

To facilitate sex separation at pupal stage, all the cocoons reaching to grainages for egg production are to be cut open either manually or mechanically using cocoon cutting machine. Cut the cocoons obliquely to avoid damage to the pupa while cutting. Manually, each skilled labor can cut about 4000–5000 seed cocoons/ day (8–9 kgs). However, it is possible to cut 80,000–90,000 seed cocoons/day (8 hours) by 4 persons by using cocoon cutting machine. Estimated damage while cutting the cocoons manually is 0.5–1.0 per cent, while in machine cutting; it is approximately 4.0–5.0 per cent . Therefore, selection of method of cutting of seed cocoons is to be decided depending on the volume of work and availability of manpower.

Table 3.2: Distinguishing Characters Between Male and Female Silkworm

Male Characteristics	*Female Characteristics*
Cocoon stage	
a) Volume: Smaller	a) Volume: Larger
b) Weight: Lighter	b) Weight: Heavier
c) Shell ratio: Higher	c) Shell ratio: Lower
Pupal stage	
a) Cuticle is slightly darker in colour	a) Cuticle is lighter in colour
b) Smaller in size	b) Comparatively bigger in size
c) Abdomen narrow and small	c) Abdomen broad and large
d) Posterior end tapering	d) Posterior end blunt
e) Presence of dot like marking in the center of ninth abdominal segment on the ventral side	e) Presence of a longitudinal slit 'X' marking on the ventral side of 8th abdominal segment extending from anterior to posterior margin of the segment.
f) Lighter in weight	f) Heavier in weight
Moth stage	
a) Usually smaller in size	a) Comparatively bigger in size.
b) More active	b) Sluggish (less active)
c) Antennae bigger	c) Antennae comparatively smaller
d) Abdomen narrow and small	d) Abdomen broad and large
e) Posterior end tapering	e) Posterior end blunt (arc like structure)
f) Caudal end with a pair of hook (Harps) like claspers	f) Caudal end with a knob like projection covered with sensory hairs

Moth Emergence

During seed cocoon preservation, pupa develops and metamorphoses gradually until it finally emerges as a silkworm moth. Moths generally emerge on 10th day of spinning in multivoltine or 12th day in bivoltine. Emergence continues for approximately 3 days. When moths are allowed to emerge in a natural way from the cocoons, a kind of digestive fluid is secreted known as 'cocoonase' which dissolve the sericin and soften the silk filament of cocoon shell to allow the moths to emerge from the shell by pushing with its head. Just after emergence, the scaly hair on the body surface of freshly emerged moths is moist, the body flexible, wings small and curled. The moth stays motionless at this stage. Soon afterwards, the scaly hair gradually dries and wings begin to spread apart. Male moths start flapping their wings continuously and crawl around in an active state while looking for female moths to mate.

Shortly before the emergence, the pupae wriggle once in a while with an audible sound. When metamorphosis is complete, the moth body and pupal skin gradually separate from each other. Before emergence, the pupal body is delicate and soft to the touch. This is the chief characteristic of this development stage. Under normal preservation conditions, exposure of light begins at 4.00 hrs in the morning and emergence essentially ceases at 8.00 hrs on the same day. Light should not be bright. Even 5 lux of light is sufficient to induce moth emergence. Trained labourers pick the emerged moths. The male moths, which emerged first, are picked out immediately to avoid mating with the females because of error in sex discrimination if any. In case copulation is observed, the copulating female moth should be discarded.

Behaviour of Male Moths

Insects chiefly employ tactile, visual, auditory and chemical means to recognize the opposite sex. In silkworm, the communication between sexes is accomplished through chemical signals. The virgin adult females after emergence emit sex attractant chemical substance, termed as 'Pheromones' or 'Social hormones' which is sufficiently volatile in nature, into the atmosphere for attracting the males. Stimulated males are anaemotactically attracted towards the source of pheromone. The responding males exhibits typical sign of alertness – first cleans its antennae, lifts the head, vibrate wings in smaller to longer amplitudes to produce a strong air current which flows parallel to the body's longitudinal axis so that it could locate the source of pheromone (Obara, 1979). He also observed that an active dancing male can draw air from a distance of 4 cm from the frontal part of its head, while wing vibrations draw air from the sides thus improving the efficiency of the antennae. Stimulated male, immensely active, runs in circles, semi-circles and zigzag paths, with distinctly bend abdomen and eventually adopts an oriented turn resulting in contact with the female genitalia. The excitement behaviour serves as an easy mode for identification of male, while females conspicuously remain inactive. In the process, characteristic dancing movements guide the males towards the direction of a female and during the dance the male could touch any part of the female body and ultimately by trial and error establishes a contact with the female genital portion and succeeds in copulation. The copulation pattern is referred to as 'end-to-end' pattern (Samson and Biram Saheb, 1994). The locking of the male genitalia with the female is adequately so firm that one has to physically slide out or pull back the male from the female while de-coupling. Hence, it is an established practice to lift the pairs by holding the female and slipping them into special enclosures (cellules) for stipulated mating duration.

Synchronization of Moth Emergence

For preparation of hybrid, the male and female moths should be essentially available simultaneously. For synchronization, the eclosion is adjusted by brushing schedules or by refrigerating the eggs/cocoons/moths. In principle, refrigeration of seed cocoons or pupae should be avoided since it has adverse effect on the overall reproductive efficiency. However, if it is inevitable, just a day prior to emergence, male and female cocoons can be refrigerated to the maximum of 7 and 3 days respectively (Jolly, 1983). The ideal and widely accepted practice is to refrigerate the male moths for a period ranging between 7 days to 2 weeks at 5°C or 7°C (Tazima, 1978, Jolly, 1983). Bivoltines are more resistant to cold storage than multivoltines. Bivoltine virgin females can be preserved at 5°C for 3 days in summer and 4–5 days in rainy season to synchronize the mating but this should not be practiced in winter season even for a day (Singh *et al.*, 1994). Cold storage should be restricted to any one stage of development to avoid deleterious effect thereafter.

Moth Collection

In advanced countries, coloured are sprayed on moths to identify the sex and breed at the time of crossing. In India, specific coloured crates are used for picking male and female moths of different breeds. The emerged moths will crawl over the newspaper through the holes and settle on the paper. The newspaper is folded and lifted and transferred to empty trays having old newspaper at the bottom. The tray should be labeled to indicate the lot number, breed, date and time of collection. The moths thus picked should be temporarily preserved in a cool and dark place to inhibit their activities, which reduces their energy consumption. When male moths smell the scent 'bombycol' secreted from the alluring glands of female moths, they will become highly stimulated which causes male moths to waste their much of the energy and even many of them may even drop to the floor from the trays which creates difficulty in the work of picking of emerged moths. Therefore, two sexes are stored in different rooms, which are kept dark. Under normal conditions, the work of moth collection should be completed by 8.30 hrs in the morning.

Moth Selection

Defective moths should be eliminated for improving the quality of silkworm eggs. Before mating, it is necessary to carefully sort and eliminate defective female moths. The major types of defective moths that are to be rejected are:

- ☆ Too long body segment, bulky body, abnormal body shape and vestigial or underdeveloped or deformed wings.
- ☆ Inert or dropped scaly hair.
- ☆ Inactive or incapable of copulation.
- ☆ Early and late emerged moths.
- ☆ Black marks or colour on abdominal segments and wings.

Only good, active and healthy moths are selected for mating. This helps in getting healthy eggs. Healthy or good eggs determine good hatching and good health of larva at the time of hatching.

Copulation

After emergence of moths are over, the mating/crossing/copulation is carried out according to the plan of hybridization.

Mating is an essential behavioural social event in the life cycle of the silkworm, for the perpetuation of population. A number of intrinsic and extrinsic factors and events of significant importance are involved in successful mating and egg deposition by an adult silk moth which besides biochemical, physiological and environmental factors also includes attraction of reproductively competent male and female moths for mating. For allowing male and female moth for mating, craft paper is spread over a wooden frame bearing the name of the breed. Generally emergence of moth occurs early in the morning. Male moths emerge first followed by female moths. When female moth emerges they take one or two minutes to spread their wings and prepare for copulation. Female moths are collected from the room and are evenly distributed on trays and then male moths are broadcasted over them and allowed to mate naturally. There should be about 20 per cent more male moths than females. After about 10–20 minutes, pairing begins naturally. After 30-40 minutes of pairing, unpaired male and female moths are picked out and gathered in a separate tray. About 300–400 females can be accommodated in a wooden tray of size 60 x 90 cm and about 400–500 male moths are broadcasted over it. The paired moths in the process of copulation are arranged evenly with adequate spacing in between to avoid interruption and decoupling. Unpaired moths are paired once again.

Sufficient time must be allowed for copulation. If it is too short, egg laying will be delayed and slow, fecundity will be less and more unfertilized eggs will be laid. On the other hand, if copulation is too long, decoupling will result in immediate oviposition and great loss. It is established by experience that pairing of 45 minutes at optimum temperature and humidity is sufficient for fertilization. However, for commercial egg production, copulation is allowed for 3–4 hrs since during this period at least two ejaculations occur. First ejaculation occurs during first 30 minutes and the second after one and half hours.

Duration of mating varies slightly with the temperature of oviposition room, silkworm breeds and number of mating to be allowed for male moths. Mating is little shorter at high temperature. It is best for the male moth to be allowed to mate once a day and should not be allowed for more than two mating. If mating is repeated on the same day, male moths should be refrigerated at 5 ± 2°C to rest for three hours between the first and second mating. Male moths should not be used for more than two times under any circumstances. If used for more than two times, it results in poor fecundity and more unfertilized eggs.

The optimum temperature during copulation should be 24–25°C for bivoltine and 27–28°C for multivoltines. The relative humidity in both the instances should be maintained at 75–85 per cent . At high temperature of 30°C and dry conditions, the moths are naturally decoupled before completion of specific duration of mating. When mating is in progress, if decoupled moths are noticed, they should be removed and left to mate a new. Pairs are to be kept in semi-dark condition.

Male Moth Preservation

Before and after pairing, male moths have to be preserved at 7°C to 9°C. The plastic trays of size 40 x 60 cm or 60 x 90 cm are usually used for preservation of male moths. Do not over crowd the male moths in preservation trays. Keep 800–1000 male moths in each tray of standard size. If male moths have to be reused on the same day for second pairing, they should be rested at least for 3 hrs at 5 ± 2°C. All trays are to be labeled to indicate the age of moths (in days) and status of moths (fresh or once mated). Preservation in refrigerator should be avoided since the temperature inside the refrigerator is not uniform and also humidity build-up besides water droplets falling on the preserved moths will render them unfit for pairing.

Decoupling

After the time set for mating is over, male and female moths are detached. This process of detaching the male and females is known as decoupling. Decoupling/de-pairing have to be done by experienced persons. Care must be taken not to apply force in pulling the moths apart, as reproductive organs could be damaged. The female moth is held between the thumb and middle finger of left hand and then pulling the male moth by right hand fingers twisting it slowly to one side. After mating, the female moth passes a considerable quantity of urine just before egg laying. The mother moth not having urinated or not having thoroughly urinated would not only have a late oviposition but also would contaminate eggs when they urinate during the course of egg laying. If urine is allowed to mix with eggs, it solidifies on the eggs and such eggs may die. In order to avoid such mortality, mother moths should be shaken to accelerate their urination. It is suggested to collect the decoupled female moths on a piece of soft cloth. Instead of in heaps, moths should be evenly dispersed on the cloth selected for the urination purpose. Then with two hands, the piece of cloth with moths on it should be lightly lifted and gently shaken up and down for 2–3 minutes. This will help the female moths to urinate thoroughly. Otherwise, the moths will delay laying their eggs and urine will spoil the craft paper and contaminate the eggs.

Oviposition

After decoupling and urination, female moths are collected and allowed to lay eggs on egg laying sheet, which are kept ready by the method described. The diversity of oviposition conditions of silkworm moth is closely related to differences of breed, duration of mating and temperature during oviposition and intensity of light. The earlier laid eggs are usually bigger than the late laid eggs in almost all the silkworm races. Also egg laying is quicker with prolonged mating and will be slow if mating duration is short. Oviposition can be accelerated, if temperature of oviposition room is high and will be slowed down if the temperature is low.

The temperature of 25 ± 1°C and humidity of 80 ± 5 per cent are maintained in the oviposition room. This is very essential for good and optimum egg production. If humidity is less, the gummy substance secreted by female moths gets dried up on the ovipositor and obstructs egg laying. This results in less number of eggs laid by the female moth. To maintain humidity, humidifiers are used in oviposition room. During summer, wet gunny cloths are hung in the oviposition room and sand beds under the tray stands are also arranged. Oviposition rooms are made dark for rapid egg laying. The doors and windows of oviposition room should be shaded with black curtains. Moths are kept in oviposition room till next morning by providing optimum environment. Usually the rate of oviposition reaches approximately 95 per cent and sometimes surpasses 95 per cent within 8 hrs of oviposition.

Moth Collection After Oviposition

Most of the silkworm eggs are laid in the same evening before 22.00 hrs. Therefore, female moths are usually collected on the following morning both in the case of multivoltine and those bivoltine, which are intended, for artificial hatching treatment. If, however, they are intended for hibernating eggs, there is no harm in postponing moth collection. However, it is being observed that most of the eggs laid on the second day of oviposition are unfertilized eggs. Moth collected for mother moth examination at later day are taken out and put into boxes indicating the date of oviposition, name of the breed, batch and serial number and stored in well-ventilated dry places. The moths should be well protected against insects or rat attacks or mould.

Male Moth Transportation

In commercial silkworm seed production centers, sometimes a critical situation arises, though not frequently, when there is shortage of male moths, despite a systematic planning. The shortage of male moths may be due to non-synchronization in the emergence of parental breeds, inadequate male population, higher pupal mortality, Uzi infestation etc. Often the weaker male moths are unable to mate successfully, due to inherent defects or improper preservation. To tide over such an exigency, one tries desperately to procure male moths from nearby Silkworm Seed Production Centers (SSPCs), otherwise precious female moths have to be rejected affecting the production and cost of production as well. For proper utilization, male moths are to be transported under ideal conditions to conserve their energy and for prudently use, demands high tools and techniques.

Harvesting of Eggs

Harvesting of silkworm eggs are carried out after moth collection. The egg sheets are checked for poor layings in case of flat card method of egg production. Poor layings are those where the numbers of eggs in a laying are much less than the standard or norms fixed for a disease free laying of a breed. These poor layings are scrapped and removed. Now the egg laying sheets are soaked in 2 per cent formalin solution for nearly 15–20 minutes. This helps in surface sterilization of eggs. These eggs are washed in water and allowed to dry in shade. The certificate of disease freeness of seed is affixed. The egg sheets must carry the name of the grainage, hybrid combination/name of the breed, date of egg laying, signature of moth examiner and number of good Dfls on the sheet. In case of multivoltines, these eggs are ready for sale. In case of bivoltines, eggs generally undergo hibernation within 48 hrs of oviposition.

Therefore, if hatching is required within 15 days, the eggs must be treated with acid within 20 hrs of egg laying. In case hatching is required after lapse of more than one month, the healthy layings after washing with formalin are sent to cold storage for preservation.

Oviposition and egg laying behaviour has been at the centre of many of the major debates. Ever since, Ehrlich and Raven (1964) report on co-evolution of butterflies and plants, Lepidopteron probably being the Texan in which the greatest number of species have been studied for some aspects of oviposition behaviour. The biological factors depending upon behavioural patterns and physiological conditions of male and female members of insects searching each other for mating and reproduction have been reported to be responsible for perpetuation of the population. Oviposition is an important physiological and behavioural aspect in the life cycle of the silkworm. It is an act of reproduction through which viable eggs are laid by fertilized female moths. Number of factors and events are involved in the successful egg deposition includes neural, hormonal, chemical, environmental, physical and behavioural. Besides these, successful and viable eggs deposition by an adult female moth also depends upon various events of significance importance of reproductive physiology *viz.*, mating, vitellogenesis, ovulation and oviposition etc.

Stimulation of Oviposition

A range of adult behaviour is exhibited in succession after eclosion in the silkworm. The eclosion occurs during the opening of a particular gate. The insects that had completed development are able to emerge at this gate. If the insects are not competent enough to emerge from pupal case, must wait for the next open gate. The term 'gate' can be applied to such event that occurs only during a limited span of time. Thus a gate is being opened for a specific period of time once each day. In female moths, the calling behaviour (pheromone release) occurred after the expansion of wings. In 1959, Karlson and Luscher suggested the name pheromone for those substances 'secreted to out side by an individual and received by another individual of the same species, in which that releases a specific reaction'. The female moth secretes sex pheromone from the lateral glands of the last abdominal segment to attract male for mating. This pheromone is known as 'Bombykol' (10–12–hexa -decadiene–

1–ol) which in real sense is a primary alcohol with a chemical structure of $CH_3 (CH_2)_2 (CH=CH)_2 (CH_2)_8 CH_2OH$.

The scales of silk moth also serve as a releaser of copulation attempt followed by wing vibration and mating dance by male moths. Mating dance helps in determining the location of the consepecific female. The chemoreceptive abilities of male to bombykol produced by female are very high. Males react to bombykol at a concentration as low as 100 molecules of attractant per cubic centimeter of air. Even a single molecule of female sex pheromone is sufficient to trigger an impulse in the male receipt cells. This is due to presence of large number of sensilla (60,000) on the bipectinate antennae of moth, whose surface area with all the remi included is approximately 29 mm^2. During mating dance, male produces a strong air current with its wings, which flows parallel to body's longitudinal axis and could trace the pheromone source by testing the air. Further dancing male can draw air from the distance of 4 cm in front of its head while wing vibrations draw air from the sides of its area, thus improving the efficiency of antennae.

The response of male moths to the female secretion pheromone is not just a simple attraction but also a complex sequential series of events by anemotaxis, which occurs, in zigzag stages as:

- ☆ Reception–antennal elevation.
- ☆ Activation–wing fluttering.
- ☆ Orientation towards the source of pheromone (anemotaxis).
- ☆ Alighting–lending in an immediate vicinity of female moth.
- ☆ Copulation–mating.

For successful oviposition, three factors or events are of utmost importance:

- ☆ The mating partner should be reproductively competent.
- ☆ The mating should be accomplished at the appropriate time.
- ☆ The mating should ensure transfer of sperm and the viscous heterogeneous secretions including male factor (oviposition stimulating substance or fecundity enhancing substance) into the reproductive tract of the adult female.

Duration and Frequency of Mating

The mating length in silkworm essentially affects silkworm seed quality and quantity. In *Bombyx mori*, natural copulation continues for 6–12 hours and sometimes because of such prolonged copulation female moth dies without laying any eggs. The number of eggs laid by a female moth depends on its genotype and development, which increases in a proportion to the duration of mating (Askari and Sharan, 1984). Punitham *et al.* (1987) stated that mating duration up to 6 hrs increased the total egg output and reduced pre-oviposition period but subsequent increase in duration resulted in negative effects.

Moreover, mating lengths of 150–180 minutes is as optimal variants for enhancing productivity and quality of silkworm seed. Further, it is to be mentioned that for production of super elite, elite and reproductive seed, mating for less than 120–150 minutes should not be allowed, while for production of commercial seed, it should not be less than 90–120 minutes. A minimum period of 45 minutes is adequate to ensure normal egg laying with regard to both fertility and number of eggs laid and two mating can easily be adopted without affecting any of the commercial traits, such as number and viability of eggs, cocoon weight, shell weight and yield of good cocoons. By resorting to repeated mating of selected males, the cost of seed cocoons and consequently the cost of seed for commercial rearing can appreciably be reduced. Petkov and Mladenov (1979) observed that male silk moth could successively mate with 8 females of the same origin. Ram and Singh (1992) stated that a single male could mate with at least 11 females in bivoltine silk moths but higher percentage of viable eggs can be obtained only from the first mating. Gupta *et al.* (1986) stated that male of multivoltine and bivoltine can mate with 16 and 18 females respectively.

The mating capacity of males of *Bombyx mori* pure breeds as compared with hybrids was studied by Benchamin *et al.* (1990) with repeated mating (6 times a day) for six consecutive days (with rest periods at 5°C between mating) and stated that fecundity in multivoltine female parent was not influenced by the male but other characters including effective rate of mating, fecundity and fertility differed significantly among different crosses. All quantitative characters seem to decrease significantly with the increase in the number of mating. However, a male moth can mate to a large number of females but many fold use of male silk moths for mating, results in decrease in the basic biological characters, which includes not only cocoon technological qualities but also the fertility of resulting moths. Further mating from 3rd fold onwards results in decrease of hatchability, viability, cocoon yield, cocoon weight, silk yield, filament length etc and hence future technologies should not allow more than two fold mating for the production of silkworm seed to obtain better results not only in fertility but also for stable cocoon crop. Moreover, it is imperative to preserve the same male moth for 3 hrs at 5 ± 2°C for successful second fold re-use.

Time of Ejaculation

After mating, ejaculation takes place at intervals of 60–90 minutes and each ejaculation is completed in about 40 minutes. The male moth does not begin ejaculation immediately after the beginning of copulation and does not separate automatically from the female until the second or third ejaculation is completed. Initially during ejaculation general semen is emitted which contains spermatozoa and consume 5–10 minutes followed by ejaculation of spermatozoa, which lasts for 10 minutes. Thereafter the emission of the semen without spermatozoa occurs again lasting for some 10 minutes (Omura, 1939). The sperm release is accomplished in a rhythmic way. When natural copulation extends to unduly longer duration, there is fear that female may die without laying any eggs.

Krishanaswami *et al.* (1973) reported that first ejaculation starts about 9 minutes after copulation and completed after 25 minutes of mating and hence optimum-mating duration of 30 minutes is adequate for obtaining desired number of viable eggs, which is sufficient to provide copulation stimulus essential for egg laying without impairing the percentage of fertility. Askari and Saran (1984) found that duration of mating has profound bearing on the recovery of eggs and increased mating duration produced higher quantity of eggs. Punitham *et al.* (1987) reported that increase in mating duration not only reduces the pre-oviposition period but also increases the egg output and enhances the hatching percentage and body weight in the succeeding generation. Duration of pairing has an inverse correlation with oviposition period.

Although in silkworm *Bombyx mori,* the virgin female moths have fully developed eggs whose development takes place during pupal stage but lack of copulation reduces the number of eggs laid. The copulation provides some essential stimulus for oviposition and in *Bombyx mori,* the stimulus that incites egg lying is not the direct stimulus of copulation but is the stimulus by spermatozoa removing from the recepticulum seminalis to the vestibulum. Drastic biochemical events occurred in the spermatophore of the silkworm after mating. This assured the presence of normal sperm and or testicular fluid in the female reproductive organ (bursa-copulatrix/ spermatheca) in sufficient number to induce the ovipositional bahaviour. On an average, normal male moth have 15.4 x 10^5 spermatozoa per head. It is generally recognized that the male factors or the oviposition–stimulating substance (OSS) derived from the male reproductive tract are transferred to female during mating and these factors have the ability to accelerate the oviposition behaviour.

The male factor has been identified chemically in some insect order including Lepidopteron and is being termed as 'fecundity enhancing substance, or 'oviposition stimulating substance (OSS)'. These factors in the silkworm *(Bombyx mori)* have been identified as 'Prostaglandin'. Oviposition is stimulated in the female moth following the transfer of prostaglandin during mating. Hence, level of prostaglandin increases in mated females. Brady (1983) stated that prostaglandin increases egg laying in the silk moth by stimulating muscles in the oviduct in the same way as is being reported in egg laying vertebrates (Wechsung and Houvenaghel, 1976).

Vitellogenesis

It is a process through which the terminal oocytes grow unto final size before ovulation. This is achieved by incorporation of female specific protein 'vitellogenin' via haemolymph. The fat body and follicle cells of the ovary synthesize the vitellogenin at a specific age in the insect life cycle. These proteins have been found to be precursor of the egg yolk protein. The vitellogenin are termed 'vitellin' when deposited in the eggs. The vitellin and other similar yolk protein make up approximately 80 per cent of the yolk proteins. The synthesis of vitellogenin is under the control of Juvenile Hormone (JH) and ecdysteroids. The juvenile hormone is secreted by Corpora Allata (CA) and is regulated (stimulated or inhibited) by neurosecretory cells of the brain and some other factors.

The ecdysone secreted from prothoracic gland and responsible for moulting, also plays a major role in reproduction (Hoffmann *et al.*, 1980) and follicle cell epithelium is the exact site of ecdysone biosynthesis. These ecdysteroids are either secreted into haemolymph or retained and accumulated in the oocytes to play a major role in vitellogenesis. During moth emergence, the brain stimulates synthesis of JH by CA. Juvenile hormone induces mating end and causes the ovary to become prepared to respond to a neurohormone, termed as 'Egg Development Neurohormone' (EDNH). EDNH triggers the synthesis and release of ecdysone from the ovary. Ecdysone, after its hydroxylation to 20–hydroxy ecdysone, stimulates vitellogenin synthesis.

Mating and Oviposition

It is well known fact that is not exactly obligatory for either ovulation or oviposition, but insemination is essential for securing a complement of fertilized eggs. It is also well established that brisk egg laying begins under the influence of mating. The stimuli, which is applied, to the female during mating are both direct (copulation, grasping of female by claspers and the repletion of bursa copulatrix with the seminal fluid) and indirect (migration of spermatozoa from bursa copulatrix to vestibulum via receptaculum seminis). Paul *et al.* (1993) and Yamaoka and Hirao (1973, 1977) observed that mating stimulates laying so definitively that within 24 hrs of mating, female laid most of the eggs stored inside the reproductive organs and stated that the mating changed the humeral condition of the female which may stimulate the motor neurons in the 9th ganglion for activating the egg laying of mated females. It is generally recognized that the male factor or the oviposition stimulating substance derived from the males reproductive glands and transferred to the females during copulation induce brisk oviposition (Fugo and Arisawa, 1992). The presence of sperm and seminal fluid inside the female reproductive organs, are prerequisites for oviposition. Oviposition rhythm or the peak of oviposition changes with duration of mating, time of mating, age of female parent and the inherent capability of the parents. Duration of mating influences the oviposition rhythm and number of eggs laid at different intervals. Longer duration of pairing showed highest percentage of eggs laid within 6 hrs of de-pairing. This is interpreted as a consequence of the interaction of factors such as increased egg maturation, successive ejaculation and optimum peak of egg laying period (Gowda, 1988).

Impact of Age on Mating Capability and Oviposition

Kovalev (1970) showed that females mated after 24 hrs of emergence laid all the eggs within 30 minutes of de-pairing. Biram Saheb *et al.* (1998) draw an inference from their studies that when one or two day's old females were utilized, which implies a significant delay in the pairing schedule, the peak of oviposition was remarkably advanced and over 70 per cent of the eggs are recovered within a short period besides a marked reduction in the retention of eggs inside the ovarian tubules. They also observed that the female moths, which were mated immediately after emergence or within 8 hrs of eclosion, deposited 81–91 per cent eggs on the first and 9–10 per cent of eggs on the 2nd day of oviposition. Quiet distinctly, the one day/two day old females upon mating for 4 hrs oviposited more than 95 per cent of eggs within 24 hrs and a small percent of eggs on the second day. Higher percentage of unfertilized eggs was reported in females mated after 6–12 hrs of emergence and it was attributed to time lapse that existed between moth emergences and mating (Gowda, 1988). Increase in age related mating schedule affects the mating potential, which ultimately causes a decrease in the protein and fat contents of the eggs. It is also stated that an increase in age of female or male moth causes a decrease in the total egg output and hatchability. Paul and Kumar (1995) reported that the mating capacity of males decreased significantly with the increase of age and established a significant negative correlation (r =–0.097) between the age of males and mating capacity. It is also established that fertility percentage decrease significantly with increasing age of females and significant negative correlation between the age of females fertility level.

Ovulation and Oviposition

Ovulation is the process of expulsion of eggs from the ovary into the oviduct and oviposition is the passage of the eggs from oviduct through external genital opening of the female to the substratum/egg laying site. Ovulation and oviposition are related at least to the extent that ovulation is a pre-requisite for ovulation. In other words, oviposition will not occur unless ovulation has taken place and in that oviposition frequently follows ovulation.

Oviposition in silkworm is dependent on a number of intrinsic and extrinsic factors *viz.*, neural, hormonal, environmental, physical, behavioural etc. It is a complex phenomenon involving multiplicity of coordinated events such as interaction between internal and external genetalia and the abdominal muscular system must be integrated in an orderly manner for systematic oviposition.

The mechanism controlling the switch from virgin to mated behaviour is not exactly known. Giebultowicz *et al.* (1990) reported that in *Lymantria* moth, the switch from virgin to mated behaviour could occur spontaneously in senescing virgin females. The same holds true also in *Bombyx mori* (Fugo and Arisawa, 1992). The bursa copulatrix and the spermatheca seems to be involved in the behavioural switching pattern from virgin to mated condition.

Mating provides a stimulus for activating ovulation. Information regarding mating is conveyed to the brain through 'spermathecal factor' released from spermatheca of mated females. This stimulation regulates the brain to release 'myotropic peptide' or 'ovulation hormone' or 'oviposition stimulating substance (OSS)', which affect the activity of the last abdominal ganglion of female moths for stimulation of oviposition. The female moths lay eggs in monolayer very compactly on the surface of the egg card. An actively ovipositing moth touches with its caudal tip the surface of the egg card or the margin of the previously laid eggs turning its abdomen from right to left or left to right. The long axis of the egg is mostly oriented radically. On the caudal tip, a pair of anal papillae is present and sensory hairs on them serves as mechanoreceptors to perceive the textural conditions of the oviposition substratum.

Temperature, Humidity and Oviposition

Considerably less is known about the effect of temperature and humidity on oviposition bahaviour. Virgin female requires optimum condition of environment for oviposition, which must be met; otherwise, it will lay only few eggs or none. Bliss (1927) was apparently the first to distinguish between the effect of temperature on the development of oocytes and its effect on oviposition behaviour. The rate of oviposition varies with temperature, which can be accelerated up to a point and then falls off rapidly. Nevertheless, the temperature limits between which oviposition can occur are often much narrower than the range of temperature over which the other activities of the same species remains normal.

The male seems often more sensitive than female to abnormal temperature. When 5th instars larvae were exposed to high temperature (32°C) for 72 hrs, the emerging males showed complete sterility. Maximum ovulation and oviposition with minimum retention of egg can occur at 25.36 ± 0.17°C (optimum) temperature and 80 ± 5 per cent relative humidity. Any fluctuations of temperature from optimum level lead to decreased ovulation, oviposition rate, fecundity and increased retention of eggs. After mating, the female moths started to look for oviposition site. Female moths start to lay their eggs mostly at dusk and about 90 per cent of the eggs that had matured are being deposited in a short span of time during night. Hardly, 9.65 per cent egg laying took place from 24 hrs to 48 hrs, while less than 1 per cent eggs in each hour are being laid from 16 hrs to 24 hrs of decoupling.

Photoperiod in insects, serves as a clock indicating the seasonal changes and influencing its life cycle, distribution and abundance. Several workers have demonstrated the influence of photoperiod on the behaviour of silkworm *(Bombyx mori)*. Photoperiod plays an important role on oviposition and in that dark condition favours rapid oviposition, while bright have opposite effect. Maximum fecundity and minimum egg retention are recorded when the female moths are exposed to fluorescent light (80 lux) during day and darkness during night.

Oviposition and Surface Texture

Texture of the substratum has an influence on the egg laying behaviour in the silk moth, which includes the number of eggs laid, total time taken for complete egg laying and the rate of egg deposition. The sensory hairs (sensory receptors) on the ovipositor and on the anal papillae in silk moth influences oviposition through the sensory input received by these hairs. Maximum numbers of eggs are reported to be oviposited on a smooth surface (Gupta *et al.*, 1990) and this decreases with the increase in roughness of the substratum.

Responses to surface topography or texture by ovipositing females are mediated by tactile setae. In *Bombyx mori*, tactile setae on a pair of anal papillae are innervated from the last abdominal ganglion. Silkworm normally lay eggs in a monolayer and the ovipositing female keeps touching the surface or margin of the group of previously laid eggs with its anal papillae turning its abdomen from side to side. If the tactile hairs on the papillae are first cut and then burnt with acid, so that dendrites are destroyed, oviposition is disorganized and the eggs are often laid in piles instead of monolayers. Besides surface topography, the plane of inclination and preservation condition of pupae also played an important role in oviposition. While slightly inclined plane of oviposition site leads to higher fecundity.

Fecundity and Weight of Pupae or Pharate Adult

The number of eggs in the body of female silk moth depends on its genotype and development. Highly significant positive correlation of fecundity in some sericigenous moths, *Antheraea mylitta, Philosamia ricini* and *Samia cynthia ricini* with pupal weight has already been established. High level of linear relationship between female pupal weight and potential fecundity as well as pupal weight and eggs laid has also been studied in various insects.

In the silkworm *(Bombyx mori)* the heaviest pupae resulted in highest fecundity. Further, the maximum larval weight, cocoon, shell and pupal weights, moth emergence and fecundity in succeeding generation were also maximum in the progeny resulted from the heaviest female pupae. Highly significant positive correlation between female pupal weight and fecundity has been demonstrated in *Bombyx mori.* Though the selection of individuals for higher fecundity depends on its pupal weight but extreme heavy weight should be avoided as it may lead to bottleneck phenomenon. Therefore, selection of moderate pupal weight should only be encouraged for eggs production as it determines increased larval weight, survival rate, filament length etc. in the successive generations. Female pupal weight, which is positively correlated with fecundity, have also been reported positively correlated with larval weight and shell ratio, cocoon weight and moth weight (Table 3.3).

Fecundity and Larval Density

Insect growth and development proceed optimally under certain population density. Larval population density has a great impact on biology, morphology and physiology of insects. Larval crowding due to inadequate rearing space has also been found to increase the larval duration and mortality; reduce the larval, pupal and imaginal weights and affects morphology, longevity, fecundity and fertility of the resulting moths in several representatives of the order Lepidoptera.

Table 3.3: Correlation Between Female Pupal Weight and Fecundity in the Silkworm *Bombyx mori*

Female Pupal Weight (gm)		*Average Fecundity*	*Correlation Coefficient*
Range	*Average*		
1.000–1.200	1.12308	443.600	+0.8720**
1.201–1.300	1.24280	468.040	+0.9209**
1.301–1.500	1.41356	487.880	+0.8646**
1.501–1.700	1 .54728	535.360	+0.8456**
1.701–1.800	1 .74440	559.280	+0.8366**
Pooled data			
1.000–1.800	1.41422	498.831	+0.7820**

** Significant at 1 per cent level of significance.

Source: Singh, 1994.

In silkworm, *Bombyx mori* the larval behaviour, yield of cocoons and cocoon features are determined by the space provided to the larvae especially during rearing of 5th instars. The space requirement for the larvae is maximum a day or two before spinning. At this stage, larvae increases by about ten thousand times in weight, seven thousand times in volume and four hundred times in body surface to become full-grown mature larvae. Under these circumstances, it becomes essential to provide adequate spacing in rearing bed to enable the larvae to eat enough actively in accordance with their growing stage to ensure successful harvest of bumper cocoon crop.

According to Hinton (1981), there exists a relation between larval density and fecundity. Further, the weight of female moth is directly related to its weight as a pupa and as a larva, and there is therefore usually a close relation between the fecundity of the adult and the weight of its pupa and larva. Positive effect of wider rearing space on larval growth, fecundity, hatchability and also on various economic characters of cocoons have been established from which it can be inferred that an increase in population density per unit area in rearing bed decreased fecundity and hatchability in resulting generations in mulberry silkworm.

Rearing space for 20,000 larvae (50 Dfls) of bivoltine silkworm *(Bombyx mori)* as adopted in different countries indicates wider spacing under Chinese schedule is better than Japanese and Indian for rearing of same number of larvae of any stage. The recommended Indian spacing for rearing of different stages is in between Chinese and Japanese. Studies conducted and comparison made for three types of spacing schedule under Indian agro-climatic conditions reveals that recommended Chinese spacing holds better than both Japanese and Indian schedules. The 'Evaluation Index' (Singh and Rao, 1993) calculated also support the view that wider spacing for rearing has positive influence on various economic parameters of seed produced. While larval density has negative correlation with fecundity and pupal weight, its duration is negatively correlated with survival (Kasivishwanathan, 1976), reelability and neatness while positively with filament length and filament weight (Ohio *et al.*, 1970) (Table 3.4).

Table 3.4: Correlation Between Various Economic Parameters in Silkworm Influencing Egg Production

Correlation Between	*Type of Correlation*
Fecundity and productivity	Negative
Fecundity and robustness	Negative
Fecundity and female pupal weight	Positive
Fecundity and moth weight	Positive
Pupal weight and larval weight	Positive
Pupal weight and shell ratio	Positive
Pupal weight and cocoon weight	Positive
Larval duration and filament length	Positive
Larval duration and filament weight	Positive
Larval duration, reelability and neatness	Negative
Larval duration and survival	Negative
Larval density and fecundity	Negative
Larval density and pupal weight	Negative

Following points are to be taken into consideration while preparing the laying for better results:

- ☆ Reproductively competent mating partner should only be allowed for perpetuation of population. The deformed, inactive, under and over sized moths should invariably be rejected.
- ☆ Mating duration should at least be 3 hrs for obtaining optimum productivity and quality of seed.
- ☆ Future technologies should not allow more than two fold mating for production of seed. Moreover, male moths should be preserved at 5 ± 2°C for at least 2 hrs for successful second fold reuse.
- ☆ Optimum temperature (25 ± 1°C) and relative humidity (80 ± 5 per cent) during oviposition should be maintained. Any fluctuation from optimum level affects ovulation, oviposition rate and fecundity.
- ☆ Dark conditions should be maintained during oviposition.
- ☆ For better egg recovery, mated female moths should be allowed to urinate by tapping/jerking/shaking.
- ☆ Texture of the ovipositing site should be somewhat smooth.
- ☆ Too much high pupal/adult weight should not be allowed for egg production but only moderate weight should be encouraged.
- ☆ Wider spacing for seed crop rearing should be followed which, have positive influence on various economic characters of paramount importance.

Unfertilized Eggs

Silk, one of the bio-resources for human life has always occupied a place of pride and there have been incessant endeavors to improve silk both its quality and quantity. Quality silkworm seed is the basic input to harvest a successful cocoon crop. Quality refers not only to the disease free nature of the layings but also to the occurrence of more viable eggs, uniform hatching, robust larvae, quality cocoons and stable cocoon crops (Singh and Saratchandra, 2004). Quality of seed is determined by many factors of significant importance wherein unfertilized eggs play an important role.

The mated female moth of the mulberry silkworm (*Bombyx mori*) lays mostly fertilized eggs (which produce viable progeny) and a very small quantity of unfertilized and dead eggs. In diapausing breeds, the fertilized eggs are pigmented and can easily be differentiated after about 48 hrs of oviposition from unfertilized eggs. Consequent to a function of fertilization, after the appearance of amnion and serosal layers, the specific pigment 'ommochrome' spreads in the serosal cuticle, which is expressed as deep/dark brown in colour (Singh *et al.*, 2002). Devoid of fertilization, apparently in the unfertilized eggs, the egg cuticle remains as such and eggs oblige to retain the same colour imparted at the time of oviposition. On the contrary, in non-diapausing, the eggs are non-pigmented due to absence of ommochrome pigment and thus the discrimination of fertilized and unfertilized eggs is rather difficult. It can be distinguished only when the eggs reach the pin-head stage in the course of development.

The unfertilized eggs remain in their original colour while the fertilized ones acquire the blue colour as a result of sclerotization of the cuticular layers of fully developed embryos. In general, under optimum conditions of management, the percentage of dead and unfertilized eggs is very low in a given population. However, it could attain significantly higher proportions when the insect is exposed to unfavorable environmental conditions for a good number of other reasons. Occurrence of unfertilized eggs or the phenomenon of sterility is however natural any living system. Though the presence of unfertilized eggs would in no way affect the fertilized ones, the frequency of their occurrence underrates the quality. Fertilized egg is a combined product of the egg cell from the female parent and the spermatozoan of the male parent. It is well known that mating is indispensable for obtaining fertilized eggs. When a normal male and female moth copulate, the male perform an ejaculation in an hour's time by discharging the seminal fluid containing two types of spermatozoa (eupyrene sperm bundles and loose apyrene sperm) accompanied by essential heterogeneous viscous secretions which are derived from the various glands of the male reproductive system consisting of glycogen, arginase, endopeptidase, exopeptidase etc. These secretions promote well-timed metabolic events for sperm maturation, motility of apyrene sperm and dissociation of eupyrene sperm bundles through lashing/flagellating movements of apyrene sperm inside the bursa-copulatrix. (Singh *et al.*, 2002). After dissociation of the eupyrene sperm bundles and consequent maturation inside the bursa-copulatrix, the sperm escape out of the spermatophore as a mass to reach ductus seminalis. Owing to the peristaltic movements exhibited by ductus seminalis, the mass of spermatozoa escapes across vestibulum and begins to pour into the spermatheca (in one hr after copulation ended). After that it gradually increases to fill the spermatheca in about 2.5 hrs after end of the copulation (Suzuki *et al.*, 1996). The spermathecal duct also involves peristaltic movements, assist sperm migration towards spermatheca (Osanai *et al.*, 1990) and these movements occur at a constant frequency after copulation (Suzuki *et al.*, 1996).

Omura (1938) reported two pathways in the spermathecal duct *i.e.* one where the sperm move from vestibulum to the spermatheca and the other from receptaculum seminis to the vestibulum, the eggs descend down from the common oviduct during the process of oviposition. As the eggs descend down from the ovarian tubules via lateral oviducts into the common oviduct and then in to the vestibulum, the spermatheca contracts releasing large number of sperms on the egg near the micropylar aperture (Narashimhanna, 1988). The sperm penetrates the ovum through the micropylar opening to accomplish further proceedings culminating into fertilization. Polyspermy is usual in mulberry silkworm. However, when there are more than two spermatozoa in one egg, only one of them participates in the fertilizing event and the superfluous ones will degenerate. Seldom, two or more sperm nuclei join resulting in the production of polyploids or dispermic androgenic individuals (Sakaguchi, 1978).

It is apparent that in some eggs, there are no nuclei, which are destined; to become unfertilized eggs *i.e.* eggs either did not receive the sperm or the sperm disintegrate inside the egg. It may be just a coincidence that when the eggs are descending down through the common oviduct into vestibulum, a few of them, unlucky, fail to receive the sperm, as the eggs stay there for a brief period of 2–3 seconds (the delay is the time required for the mother moth to verify the substratum in all the three dimensions through its sensory hairs present on the anal papillae in order to forward a signal about the availability of clear space for the release of eggs) as they are pushed down by descending eggs and hence they become unfertilized eggs. Thus, any egg contained in the ovarian tubules could transcend down to become an unfertilized egg.

The sperm, which has penetrated into the egg, remains lodged in the anterior part of the egg until the completion of the egg pro-nucleus. In the mean time, the tail of the sperm separates from the head, near the top of the egg a centrosome appears and an aster is formed. The sperm head gradually swells into a spherical body, forming the male pro-nucleus. As soon as the maturation division is completed, the male and female pro-nuclei approach and finally fuse accomplishing the vital function of fertilization. They unite to form the zygote with twice as many chromosomes and thus creating a fertilized eggs.

Causes of Occurrence of Unfertilized Eggs

In the egg, which has just been laid, there is only yolk filling it, but the serosa/amnion nor the chromosomes can be seen. It is only subsequent to fertilization that the cleavage is initiated signaling morphogenesis followed by the formation of blastoderm, germ band and the appearance of the embryonic layer, namely the amnion and serosa. However, in the unfertilized egg, either the sperm is missing or egg nucleus/sperm are not adequately efficient to participate. Thus, none of these developments can be initiated and no morphogenesis proceeds and eventually the unfertilized eggs collapse to death in course of time and hence obviously no viable progeny emerge out. According to Toyama (1909) (as quoted by Tanaka, 1964), the unfertilized eggs are characterized by:

- ✰ No change in egg nucleus and yolk or
- ✰ Changes in colour a little but nuclei do not undergo any division and yolk separates to irregular yolk balls or
- ✰ Egg nucleus divide and produce many yolk nuclei but do not form germ band or even sometimes form imperfect germ band but do not develop further.

Kovalev (1970) stated that unfertilized eggs are the result of some abnormality in the sexual organ of the male and female or abnormal development of micropylar end of the egg. The unfertilized eggs arise besides due to unfavorable environment during the period of spinning, cocoon preservation, imperfect handling of moths during mating and oviposition, cold storing of pupae/moths and indiscriminate use of male moths etc. It is also reported that the unfertilized eggs are found in large numbers among the eggs laid at the end of oviposition as compared to the eggs laid at the beginning of oviposition by the same female moth (Ming, 1994). The various possibilities on the occurrence of unfertilized eggs are discussed briefly as under:

- ☆ Temperature is one of the most important abiotic factors with pervasive effect and plays a vital role in the physiology of silkworms. More so, the silkworm represents poikilothermic species, most of whose physiological processes are governed by temperature. Wigglesworth (1972) stated that temperature range between which reproduction occurs are often much narrower than that in which the other activities of the same species remain normal. Reproduction as well as development of an insect is temperature dependent and hence, it immensely influences the dynamics of insect population. Thus, the various operations pertaining to seed production are performed within optimum limits of temperature to avoid reduction in egg yield and fertility rate (Kamble, 1997).
- ☆ Both low and high temperature during silkworm rearing, cocoon preservation or egg production causes unfertilized eggs (Yokoyama, 1963). The impact of higher temperature is more intense and prolific than the lower one. Number of investigations points to the fact that exposure of male silkworms at higher temperature from the time of spinning to pre-pupal period (32–33°C) brings male sterility (Sugai and Kiguchi, 1967; Sugai and Hanaoka, 1972; Sugai and Takahashi, 1981; Das *et al.*, 1996) and results in increase in the occurrence of unfertilized eggs (Jolly, 1983; Narashimhanna, 1988; Biram Saheb and Gowda, 1987; Ming, 1994). The pre-pupal stage *i.e.* spinning to pupation is the most sensitive stage of the male silkworm to have sterilizing effect at high temperature followed by the pupal period. It is interesting to observe that high temperature does not inflict the sterilizing effect in the larval period from hatching to the middle of the fifth instar (Sugai and Hanaoka, 1972). Even the seed cocoons when exposed to high temperature of 35°C or in combination with low relative humidity, causes decline in the egg recovery besides corresponding increase in the incidence of

unfertilized eggs (Ayuzawa *et al.*, 1972) with more than 90 per cent of the eggs laid at 35°C remain unfertilized. Lowering the relative humidity further increases the percentage of unfertilized eggs at higher temperature (Gowda, 1988). He further stated that when freshly formed silkworm pupae were exposed to 35°C till oviposition, only 31-42 per cent of moths laid normal eggs implying a serious damage to the physiology of pupae.

- ☆ Higher temperature at the time of egg layings also significantly promotes the occurrence of unfertilized eggs (Kovalev, 1970). Similarly, lower temperature too enhances the frequency of unfertilized eggs (Ayuzawa *et al.*, 1972). The elevation of temperature tends to enhance the rate of development and interferes in various genetic and physiological functions, so either the regular developmental pathways deviate so much that there is no emergence at all or when there is emergence, the moths of either sexes get partially or completely sterilized with the net result of production of high degree of unfertilized eggs (Gowda, 1988).
- ☆ The copulatory organs of the males treated with high temperature at pre-pupal stage did not exhibit any visible abnormality (Sugai and Hanaoka, 1972). Sugai and Takahashi (1981) after morphological studies of the male copulatory organs reported marked reduction in the production of apyrene sperms with a little change in the quantum of eupyrene sperms. Eupyrene sperms are formed at the end of the larval period till the beginning of the pupal period while the apyrenes are produced only in the pupal period but still high temperature in the pharate pupal stage affected the early stage of formation of apyrenes (Sugai and Takahashi, 1981). High temperature treatment causes marked abnormalities in apyrenes and many of them remain in the testicular chamber without penetrating the basement membrane. Silk moths normally produce about twice as many apyrenes as eupyrenes and the ratio of eupyrene and apyrene is 1:1.84 at 25°C while at higher temperature (32°C), it is 1:0.85 showing that the production of apyrenes was markedly reduced at 32°C during pharate-pupal stage (Osanai *et al.*, 1989). Moreover,

they further stated that in the moths treated at 32°C, the number of eupyrene spermatozoa was only 67.2 per cent and apyrenes spermatozoa was 31.2 per cent of that moths preserved at 25°C. Thus, the production of both types of spermatozoa and especially apyrenes are very much affected in the moths treated at 32°C than in those of 25°C exposed moths (Osanai *et al.*, 1989). Hence, the occurrence of unfertilized eggs is comparatively higher during the hot summer months compared to that in other seasons.

- The apyrene play an exceedingly important and a beneficial role. Osanai *et al.* (1987) observed that the apyrenes promote dissociation of the eupyrenes bundles within the spermatophore through their vigorously flagellating movements possibly through the digestion of the prostatic secretion or an endopeptidase. Thus, though a large number of eupyrenes are ejaculated into the bursa-copulatrix of the female moth copulated with the sterile male moth, as the apyrenes quantity is low, the dissociation of eupyrene bundles is not efficient and hence neither eupyrenes nor apyrenes are detectable in the spermatheca. However, when male sterility is partial, significant quantity of eupyrenes and apyrenes prevail and that they also move from the ductus seminalis to vestibulum and to spermatheca and compete for entering the eggs for participating in fertilization. This reveals that the occurrence of unfertilized eggs may be due to either complete breakdown of spermatogenesis, inadequate production or failure to transport the sperms effectively by the male into the genital tract of the female and incapability of sperm movement from bursa-copulatrix to spermatheca or partial movement of sperm to sperm store house. Thus, a reduction in the production of sperm seems to be the main causative reason for male sterility rather than complete cessation in the production of sperms.
- When the female pupae or moths were exposed to high temperature of 35°C and crossed with normal males, the egg yield was remarkably reduced. The majority of eggs laid comprised of unfertilized eggs signifying that even in the females, the reproductive function is greatly disturbed and the reproductive organs may fail to transport the

sperm to spermatheca. It is also possible that the egg nucleus lost its identity and become unfit for fertilization or disintegrated. High temperature interferes with the developmental process and cause physiological breakdown leading to partial or complete loss of reproductive potential in either sex (Gowda, 1988).

- The effect of low temperature on insects is manifold. In case, the temperature during seed cocoon preservation is lower than 20°C, unfertilized eggs will increase and further consolidate into higher proportion at 15°C (Kovalev, 1970; Tazima, 1978; Ming, 1994). For the purpose of synchronized emergence, seed cocoons are cold stored for 3–7 days only. Cold storing of seed cocoons besides lowering the percentage of moth emergence, egg recovery and fertility would also enhance the percentage of unfertilized eggs (Ming, 1994; Christiana *et al.*, 2003) even to the extent of 20 per cent .
- Chilling of female moths is yet another cause for the occurrence of unfertilized eggs. Chilling at 5°C should be limited to only 2–3 days as unfertilized eggs are produced only if it prolongs further (Tazima, 1978). Also observed that male moth can withstand refrigeration for 2 weeks but too long refrigeration causes unfertilized eggs. Contrary to this, Ming (1994) observed that cold storing of male moths beyond 4–5 days affected the mating capability besides enhancing the frequency of unfertilized eggs. Gowda (1988) reported that the incidence of unfertilized eggs increases with the increase of the duration of cold storing of male moths.
- When the mated female moths are cold stored beyond 4 days, the moths lay unfertilized eggs that increase with the increase in the duration of the cold storage. The increase in the number of unfertilized eggs might be due to the inactivation of sperms present in the spermatheca of the cold stored female moth (Raju *et al.*, 1990). It is also possible that the process of ovulation is so much accelerated that most of the eggs briskly or swiftly escapes out of the vestibulum and hence become unfertilized. Failure to transport the sperms effectively inside the female

genital tract and the entire consignment or portion of sperm from the spermatophore may be retained inside the bursa-copulatrix and may not reach the spermatheca due to some defect in the bursa-copulatrix or ductus-seminalis or vestibulum. The organs associated with the passage of sperm *i.e.* ductus seminalis; vestibulum and recepticulum seminis are severely affected and may not fully encourage the transport pathway.

☆ Mating is another important event that if handled inadequately, leads to significant increase in the number of unfertilized eggs. Mating should be conducted at optimum temperature for better recovery and higher or lower temperature range will produce fewer eggs with higher percentage of unfertilized eggs (Ayuzawa *et al.*, 1972). Singh *et al.* (2003) reported that mating at higher (35°C) or lower (20°C) temperature in combination with optimum relative humidity do not influence the fertility status of male or female and the recovery of eggs and the increase in the number of unfertilized eggs. However, when oviposition and or pairing of moths are conducted above 35°C under optimum humidity conditions, fecundity and fertility were greatly reduced and the number of eggs retained inside the body of the moth increased proportionately.

☆ The success of mating depends on the overall physiological, reproductive and health status of the participating male and female moths, their age, time and duration of mating and the number of times the male moth are used for mating (Singh and Tripathi, 1995; Singh and Saratchandra, 2004). Sufficient time should be allowed for copulation and if it is too short, egg layings will be late and slow and fewer eggs will be laid with many of them being unfertilized (Ming, 1994). Jolly *et al.* (1964) and Singh and Tripathi (1995) observed that lower mating durations caused sterility and opined that 3–4 hrs of copulation time should be allowed so that at least two ejaculations from the males are exploited. Mere physical union of male and female moths would not suffice unless there is ejaculation of sufficient quantity of spermatozoa and other heterogeneous viscous secretions into the female bursa-copulatrix including the Oviposition Stimulating Substance (OSS) (Biram Saheb *et al.*, 2005).

- Indiscriminate use of male moths is yet another cause of occurrence of unfertilized eggs. Continuous mating without rest between two mating affects fecundity and increases the unfertilized and dead eggs (Subramanyam and Narasimhamurthy, 1987; Biram Saheb and Gowda, 1987; Sreenivasababu, 1993). It is well known that when a sterile male moth is crossed to a normal female; the eggs recovered are unfertilized eggs (Fugo and Arisawa, 1992). Also, there is increase in the incidence of unfertilized eggs when the female moth is affected with pebrine disease (Jolly, 1983).

Strategies to Avoid Unfertilized Eggs

- Maintain 23–24°C temperature during final age seed crop rearing.
- Maintain optimum temperature (25 ± 1°C) and humidity (80 ± 5 per cent) during seed preservation and oviposition.
- Proper photoperiod during pre-eclosion and eclosion period is to be provided.
- Ensure proper and perfect handling of moths during eclosion, mating and oviposition.
- Conduct silkworm rearing by adopting recommended package of practices of seed crop rearing.
- Duration and frequency of mating, age of moth on mating, time of mating, re-use of male moth for successive mating, multiple mating etc. should be judiciously decided to harvest quality seed.
- Indiscriminate use of male moth should be avoided.
- Only healthy moths are to be selected for pairing.
- Surface texture of the substratum should be somewhat smooth for egg layings to avoid delay for the release of descending eggs.
- Cold storing of seed cocoons, moths and eggs should be avoided. If at all necessary for synchronization, it should be for a minimum recommended period only.

Methods of Egg Production

The success of sericulture industry depends on quality of silkworm eggs. At the egg production centers, the eggs are collected with the aim of:

- ☆ Producing heterozygotic hybrids for higher productivity during commercial rearing and stable cocoon crop.
- ☆ Producing pure eggs that can be used to produce seed of parent variety.

The process of egg collection in these two methods is completely different. In the case of collection of eggs meant for utilizing eggs for P2 and parent variety, each and every moth is tested microscopically to prevent transovarian transmission of pebrine infection. In case of commercial egg production, meant for producing cocoons for reeling, mass mother moth examination is carried out. Therefore, the method of egg collection can be divided into two categories:

1. Segregated egg laying
2. Mixed egg laying

1. Segregated Egg Laying

The segregated egg laying method is performed in production of grand parent eggs of F1 hybrids and of parent breeds where each and every moth is tested for pebrine disease. Segregated egg laying method is sub-divided into two groups:

(*i*) Pasteur method

(*ii*) Cellular method

Pasteur Method

This method is adopted for the production of parental silkworm breeds and grandparent silkworms of F1 hybrids. In this method, one thick sheet of egg laying paper of size 32 x 18 cm^2 is divided into 28 partitions. In each partition, one moth is allowed to lay eggs. Enough space is provided between two egg laying moths with the help of cellule/rings to prevent the mixing. After oviposition the moths are transferred to another moth box with the same number of partitions labeled with their serial numbers for microscopic examination. This method is convenient for individual mother moth examination and is less expensive compared to cellular bag method, convenient to eliminate inferior layings but in this method large quantities of thick sheet of papers are required besides being more labourious.

Cellular Method

This method was also devised by Pasteur in France with the idea of avoiding pebrine infection and hence belongs to a type of segregated egg layings. In this method, small cellular bags of thin cloth or paraffin paper perforated with small holes in which female moths are allowed to lay eggs individually are used. Female moths after copulation is kept in this bag individually and tied. After oviposition, the moths are subjected for microscopical examination. In this method, there is no confusion about the female moth and also the numbers of eggs laid are more, because moths are left free besides perfect pebrine examination. Moreover, this method is even more laborious than Pasteur method, time consuming, more expensive and requires lot of space. Therefore, this method is not popular.

2. Mixed Egg Laying

This is also known as collective moth egg collection method and is adopted for production of commercial F1 hybrid silkworm eggs. Mixed egg laying method is sub-divided into two groups:

(*i*) Flat card method

(*ii*) Loose formed method

Flat Card Method

In this method, certain number of moths is permitted to lay eggs in a definite area of a sheet of craft paper. Generally, the egg laying area is about 400 cm^2 (20 x 20 cm) but somewhat different in moth number for laying eggs according to varieties, fecundity capacity of moths, productive seasons, rearing environment, quality of seed cocoons etc. The principal is that the area is full and flat without overlapping of eggs. This method is simple, convenient and labor saving but does not permit individual selection or elimination of inferior layings. After oviposition, all the moths are collected and put together for moth examination. It is more convenient in washing, disinfections, incubation and brushing/collecting newly hatched larvae. This method can be practiced only in places where pebrine is not prevalent because if even one moth is affected with pebrine in a sheet, the entire egg card is rejected. This method is popular for collecting eggs to be diverted for commercial rearing.

Loose Formed Method

Loose eggs are free eggs not attached to the sheet and the production of such eggs by detaching them from the sheet is known as loose egg production. For production of commercial silkworm eggs, the loose-formed method is a more advanced method in advanced silk producing countries. In Europe this method is obligatory because of varietals characteristics of the breed. In Japan, all silkworm eggs for commercial rearing are produced employing this method only. This method is commonly used in Eastern Provinces of China also. But, in India the concept of loose egg production has not percolated to the desired magnitude. Experiences have shown that, the reasons for non-acceptance or reluctance from both the egg producers and farmers are basically due to lack of technical know-how and inadequate infra-structural facilities and difficulties encountered during brushing. The loose eggs are packed in loose egg boxes. For preservation of loose eggs, either starched cotton cloth or starched thick craft paper is selected so that they could be used repeatedly and make loose egg production more economical. Following equipments are required for the preparation and production of loose eggs (Table 3.5).

Generally, 40–50 gm of arrowroot/starch is required to prepare one liter of paste in which 10–15 gm of boric acid is added while heating to prevent mould attack. One liter of paste is sufficient to smear 20 m^2 surface areas of sheets. For preparing the paste, first starch is weighed and dissolved in a little quantity of water and kept separately. Rest of water is heated till its boiling and the prepared solution is then added to this boiling water and stirred well and boiled till it becomes sticky. The paste is smeared on the egg sheet to create a thin film over it. Starch smeared sheets are dried under shade. Size of egg sheet varies according to the size of the tray as oviposition is allowed in trays for convenience. Size of egg sheet should slightly be more than the size of the tray. Egg sheet of size 105 x 75 cm is more ideal to befit 90 cm x 60 cm oviposition tray. For egg laying, spread the starch coated egg sheet in the oviposition tray in such a way so as to cover the entire floor surface area besides four sides of the tray.

Table 3.5: Equipments Required for Preparation of Loose Eggs

Sl.No.	*Equipments/Materials*	*Purpose*
1	Craft paper sheets	Oviposition of moths.
2	Starch (Arrowroot powder/ Maida)	Smearing on the sheets.
3	Oviposition trays	Oviposition of moths.
4	Oviposition stands	Keeping oviposition trays.
5	Egg washing trays	Washing and collection of eggs.
6	Nylon mesh bags/perforated containers	Collection of eggs and Acid treatment.
7	Acid treatment equipments	Acid treatment.
8	Egg drying unit	Drying of eggs.
9	Loose egg preservation trays	Egg preservation.
10	Winnowing unit	Elimination of dead/bad eggs.
11	Electronic top loading balance	Weighment of loose eggs.
12	Loose egg boxes/cases	Packing of eggs.
13	Loose egg brushing frame	Incubation and brushing of eggs.

After emergence, male and female moths are allowed for copulation for specified period of time and after de-pairing, female moths are broadcasted uniformly in the oviposition tray. The density of moth broadcasted per sheet varies according to the strain of silkworm and its voltinism. Under Indian conditions, approximately 250–275 of bivoltine or 275–300 of multivoltine moths can be distributed in 0.50 sq. meter base area. The trays are kept in oviposition room under dark conditions at optimum temperature (25°C) and humidity (80–85 per cent) for good laying recovery. The female moths are allowed to lay eggs for 24 hrs in the dark room on the starch paper sheet. After egg laying, on the second day of oviposition, sample of mother moths is collected for examination. Sheets are then collected and soaked in cold water for about 20–30 minutes. As a result, the gummy substance of the starch dissolved in water. The eggs become loose now. The eggs are removed from the sheet by hand or soft brush gently. The removed eggs are collected into nylon bag (doubled layered–35 cm x 25 cm) with hand loops tied at the noose of the washing tray and washed in water to remove starch. Each bag will hold 700–1000 gm of eggs released from 12–15 sheets. Loose eggs after collection in nylon bag have to be washed in running water for 5 minutes. To avoid clustering of eggs, 0.2–0.3 per cent freshly prepared bleaching powder solution is used while washing. Concentration of bleaching powder exceeding 0.5 per cent should be avoided as it is detrimental and has ovicidal effect on silkworm eggs. Use fresh solution of 10 liters for every bag of 750–1000 gm of eggs. The treatment not only helps in making the grains free but also serves as a surface disinfectant to the eggs. After treatment, the eggs are thoroughly washed in water to eliminate the traces of chemical, if any. The eggs are then collected into nylon mesh bags. The nylon bags are gently splashed to remove the excessive water contents. Water temperature during washing and collection of eggs should be in the range of 20–25°C.

For the removal of undesirable eggs *viz.*, unfertilized and dead eggs, salt solution of specific gravity 1.06–1.09 is prepared. Eggs are first dipped in 1.06 specific gravity salt-water solutions wherein unfertile eggs float on the surface while fertile and dead eggs sink to the bottom. The floating unfertile eggs are removed and eggs from the bottom are transferred to 1.09 specific gravity salt water solution. Hence, the fertile eggs float on the surface of the salt solution while dead eggs will sink to the bottom. The floating fertile eggs are collected and the sunken dead eggs are rejected. The fertile eggs are again washed in water to remove salt. The washed eggs are dried in shade and if hatching is required within 10–15 days in bivoltine, perform acid treatment as per schedule. For elimination of lighter eggs, feed the eggs into hopper of the winnowing unit and switch on the fan, discard the lighter eggs, which are blown away. Winnowing of hibernated eggs eliminate the unfertilized eggs satisfactorily. Regulate the speed of the fan of winnowing machine as per requirement for every lot winnowed. These eggs are ready for packing. The eggs are packed volumetrically.

The weight of eggs varies from breed to breed, season to season, crop to crop and even day to day during incubation. A unit of 25,000 eggs equivalent to 50 Dfls @ 500 eggs per laying is considered as a standard unit. The box carrying loose eggs are made up of light wood frame, of which both sides are packed with thin cloth. On one side of the cloth at the top corner, a slit is made through which eggs are poured and slit is closed and sealed with sticker and an index slip is affixed on the frame. The index slip must indicate name of the grainage, lot number, breeds packed, quantity, laid on date, expected released on date, probable date of hatching etc. Before filling the eggs in the boxes, the standard weight of silkworm eggs is calculated. Bivoltine eggs generally weigh between 1500–1700 and multivoltine 1800–2100 eggs per gram. The eggs could be weighed and packed according to the weight. For this, weigh one gram of eggs on the electronic top loading balance and count them. Then calculate the weight of eggs to be packed *i.e.* weight of eggs to be packed = 25000 ÷ Number of eggs per gram *e.g.* if number of eggs per gram = 1600, then weight of eggs to be packed = 25,000 ÷1600= 15.62 to 17.0 gm/box.

Alternatively, instead of weighing every time, after determining the weight of 25,000 eggs, the weighed eggs can be filled in a narrow test tube and the level can be marked with a glass marking pencil. The same volume may be used for subsequent measurement for packing of eggs. Since, weight loss is observed with every day's development, the counting and packing should be done simultaneously on the day of packing. Preserve the boxes horizontally in ideal conditions.

Incubation of Loose Eggs

Incubation is the first step for achieving a successful cocoon crop. Eggs transported in the loose egg boxes are transferred to incubation-cum-brushing frames (36 cm x 23 cm x 1.5 cm). The eggs are spread uniformly on the frames. The eggs are covered with tissue paper and incubation frames are piled-up in the incubation room wherein 16 hrs light, 8 hrs dark, 25°C temperature and 80–85 per cent humidity are provided till the head pigmentation day/stage. Light intensity on egg surface to be above 10 Lux.

Brushing of Loose Eggs

When the eggs reach the head pigmentation stage, the loose eggs are removed from the egg boxes and spread on a tissue paper. Black boxing is carried out as per procedure for 48 hrs. The eggs are exposed to light after 48 hrs of black boxing to get more than 90 per cent hatching. On the day of stipulated hatching, at the time of exposure to light, two mosquito nets are spread over the loose eggs one over the other. Leaf is sprinkled over the net and the larvae crawl onto the net. The first net is lifted and placed onto the rearing tray and larvae are brushed using a soft feather. The second net facilitates separation of hatched egg shells and the unfertilized eggs.

Advantages of Loose Egg Production

Eggs in loose forms are produced to ensure:

- ☆ Supply of quality silkworm eggs with higher hatchability.
- ☆ Supply of uniform and assured quantity of eggs independent of breed/hybrid/season/region/source of its production.

Production and distribution of loose eggs has many advantages both qualitatively and quantitatively, some of which are presented below:

- ☆ Facilitates elimination of unfertilized, dead, poor quality and defective eggs to ensure supply of only good quality eggs and hence increased hatchability and cocoon productivity.
- ☆ The egg recovery in grainages could be substantially increased as the healthy laying with low fecundity is also collected.
- ☆ Surface sterilization of eggs can be accomplished more effectively and completely.
- ☆ As loose eggs are supplied in standard unit by weight, distribution is considered more perfect as each and every indenter have equal number of eggs for same quantity of Dfls. A unit number of 25,000 eggs are packed in each loose egg case which is considered equivalent to 50 Dfls on paper-card.
- ☆ Since there is uniformity in distribution of eggs, evaluation/comparison between different breeds/hybrids and seasons on the performance of characters are more accurate and scientific.
- ☆ Handling is easy and space required during egg laying by moth is comparatively less.
- ☆ Acid treatment of bivoltine silkworm eggs (pure as well as hybrid), their preservation, incubation and transportation is easy and more convenient.
- ☆ Quality of eggs can be judged more accurately.

- ☆ Economical seed production process.
- ☆ Easy for mass egg production.

Inspite of all these advantages, the production of loose eggs is still not very much popular in India at commercial level due to certain disadvantages like:

- ☆ Calculation of hatching percentage is not accurate and at the same time difficult and time consuming also.
- ☆ This method cannot be advocated for reproductive egg production.
- ☆ Technique for separation of unfertilized eggs is still not very accurate.
- ☆ Loose egg production is more laborious and time consuming if performed in smaller quantities.
- ☆ Since the minimum unit handled in each case is 50 Dfls, farmers have to inevitably take the eggs in multiples of 50 Dfls irrespective of their rearing capacity, which sometimes leads shortage of leaf at the later stage of rearing leading to poor cocoon crop.
- ☆ Surface disinfections of eggs though carried out at the time of production but cannot be repeated at chawki rearing centers before incubation.
- ☆ Handling of loose eggs is cumbersome as differentiation of non-hibernating and hibernating eggs is difficult.
- ☆ Lack of expertise in loose eggs preparation and insufficient infrastructure also comes in the way.
- ☆ Lack of users friendly suitable equipments for egg preparation.

Major differences between flat card/sheet egg and loose egg production technology are furnished below in brief (Table 3.6).

Old Method of Eggs Production

There are two other methods, which were used in the production of eggs in the past but are not in use at present. These methods are -

Industrial Method

In this method, 50–100 cocoons from the lot is selected and preserved at 30°C temperature and 70 per cent humidity. The moths

Table 3.6: Differences Between Flat and Loose Egg Production Method

Flat Card/Sheet Egg Production	*Loose Egg Production*
(*a*) Ordinary craft sheets without starch coating are used for egg laying.	(*a*) Ordinary craft sheet coated with starch are used for egg laying.
(*b*) Craft sheet used for egg laying is mostly 22.5 x 28 cm in size.	(*b*) Starch coated craft sheet used for egg laying is of 75 x 105 cm in size.
(*c*) On each sheet 20 moths are left to lay eggs.	(*c*) On each sheet 250–300 moths are left to lay eggs.
(*d*) Cellules are placed over the moth to restrict the egg laying area.	(*d*) Cellules are not used but moths are spread uniformly on the sheet coated with starch.
(*e*) Eggs are supplied after disinfections and acid treatment (Bivoltine).	(*e*) Eggs laid on sheets are loosened, surface disinfected, acid treated, quantity measured, packed in loose egg boxes and supplied.
(*f*) Followed for P3, P2 and P1 seed production and also for F1 commercial seed production.	(*f*) Followed only for F1 commercial seed production.

emerged are examined for pebrine disease. If pebrine infection is not noticed, the moths are allowed to lay eggs on card in groups. In this case, the moths are not subjected further to microscopical examination. This method is very simple, requires less equipment and labor besides saving time and thus reducing cost of production of laying. But this method is not popular because disease laying may be passed to the rearers. This method can be practiced only in areas where pebrine is not prevalent.

Biological Method of Eggs Production

Prof. E.F. Povakov advocated biological method of egg production. As soon as larvae start pupating, they are subjected to high temperature of 33.8°C in a special compartment for 16 hrs a day at 55–65 per cent humidity. High temperature reduces the chances of pebrine by activating phagocytes in the body of pupae. Moths emerged are allowed to lay eggs as per standard procedure. These moths are not subjected for microscopical examination for diseases. But this method has many disadvantages *viz.*, more number of unfertilized and dead eggs are produced, percentage of emergence is reduced due to high temperature treatment and also causes increased occurrence of defective moths in the treated batches and hence not in-vogue.

Relatively recently, with the objective of increasing efficiency and to reduce drudgery in grainage operation following eco-users-friendly equipments have been innovated, developed, and or improved with the advantage of operational convenience, user friendly operations, durability besides cost-effectiveness.

- ☆ Seed cocoon transportation crates
- ☆ Seed cocoon/pupae preservation tray (plastic)
- ☆ Seed cocoon de-flossing machine
- ☆ Seed cocoon/pupae preservation stand
- ☆ Oviposition stand, Incinerators for waste disposal
- ☆ Circumferential room heater
- ☆ Humidifier, Automatic lighting system
- ☆ Male moth preservation tray (plastic)
- ☆ Moth preservation chamber
- ☆ Moth crushing machine, Cyclomixer

- ☆ Loose egg washing tray
- ☆ Loose egg drying chamber
- ☆ Loose egg winnowing machine
- ☆ Improved acid treatment bath
- ☆ Egg separation funnel
- ☆ Loose egg cases
- ☆ Incubation and brushing frame
- ☆ Corrugated sheets, Composite cellule
- ☆ Moth picking nets.

Chapter 4
EMBRYONIC DEVELOPMENT

The structure of egg is the determiner of the pattern of development. The structure and development of eggs in various insects have been described by many workers. The first detailed study on the development (embryogenesis) of silkworm, *Bombyx mori* egg was carried out by Toyama. Since then embryological studies were confined especially in solving the practical problems, such as identification of suitable stages for refrigeration of early embryos and the initiation, continuation, maintenance and termination of diapause in order to develop an effective and efficient model for long term cold storage of silkworm eggs. Takami (1969) was apparently the first who published a review on the embryogenesis of the silkworm, *Bombyx mori*. During these studies, morphological changes occurring during different embryonic stages in the silkworm were described. In these studies, morphological changes of embryos were used to assess their long term survival. Moreover, the tolerance of eggs to cold storage varies with the stage of embryonic development besides genotypes because of adaptation phenomenon, metabolic changes and also genetic variations etc. (Sander *et al.*, 1985, Sonobe *et al.*, 1986) and hence the recent studies on silkworm eggs becomes more concerned with physiology, biochemistry and metabolic activity associated with termination of diapause (Yaginuma *et al.*, 1990; Yaginuma and Yamashita, 1991; Singh *et al.*, 2002; Singh and Saratchandra, 2004) for effective handling.

I. Structure of Egg

Shape, Size and Colour

Takami (1946) reported that eggs of the silkworm (*Bombyx mori*) belong to non-regulative type. It is short, spheroid, slightly attenuated at the anterior end where the micropyle is situated and laterally flattened. The shape of egg is entirely dependent on the shape of chorion, which is formed in maternal body. Hence, the egg shape is inherited pseudo-maternally. Though, the size and weight of eggs varies from breed to breed, voltinism to voltinism and even season to season but it is approximately 1.30 mm long, 1.00 mm wide and 0.6 mm thick (Miya, 1984) having 0.5 mg weight and about 1.075 specific gravity. Average size of eggs of different types of breeds is presented in Table 4.1. Among the eggs of a moth, the ones which are being laid first are heavier than those laid later. The egg weight changes due to environmental conditions and even the eggs of same genotype have weight variations due to differences in the rearing conditions. The average major axis and minor axis of silkworm eggs are 1.22 ± 0.03 mm and 0.98 ± 0.02 mm respectively (Yukata, 1996).

Table 4.1: Average Size of Eggs of Different Types of Breeds

Sl.No.	*Types of Breeds*	*Length of Egg (mm)*	*Width of Egg (mm)*
1.	Japanese	1.30	1.21
2.	Chinese	1.29	1.21
3.	European	1.43	1.27
4.	Indian (Mysore)	1.05	0.87

Except few exceptions (white and red eggs), the eggs (*Bombyx mori*) are usually brownish yellow in colour. Generally deep coloured eggs are common in Japanese breeds, while eggs of Chinese breeds are commonly light in colour. Usually, the chorion is colourless and semitransparent and hence the yolk colour is easily observable through shell in newly laid eggs. Yolk is normally light yellow in colour but in some genotypes, it is dark yellow. The colour of yolk is entirely dependent on the maternal genotype, having a close relation to the haemolymph and cocoon colour of the mother moth (Tazima, 1964).

The colour of egg is also due to the colour of serosal cells. The pigment present in serosal cells is attributed to one of the amino acids–tryptophane. The pigment is formed in a series of steps in a sequence: tryptophane → oxytryptophane → formylkynurenine → kynurenine → 3-hydroxykynurenine → chromogen → chrome (pigment). In this series of reaction, certain enzymes participate. The eggs in which pigment is not formed, indicates absence of enzymes resulting in white eggs. The eggs receive gluey substance of moth on the underside during deposition to adhere to the substratum/egg card. The gluey substance is being secreted from the sebific gland or accessory gland, opening at the vestibulum.

Eggshell (Chorion)

A new generation begins at the moment of fertilization of an oocyte by an appropriate sperm. After fertilization, a series of well-defined structural and biochemical changes occurs within the oocyte leading to genesis of offspring (Margaritis, 1985). For these processes to occur successfully, the egg has to be enclosed in such an entity as to permit embryogenesis to proceed safely. The major role played in this direction is by an oocyte covering, which is commonly referred as eggshell. The physiological function of eggshell is:

- ☆ To allow and facilitate sperm penetration into the egg for fertilization which will lead to genesis of offspring.
- ☆ To protect oocyte from unfavorable environmental conditions such as extreme temperatures, high humidity, dryness (Yamashita and Hasegawa, 1985) and pathogenic attacks (Hinton, 1981).
- ☆ To ensure adequate oxygen supply for all biochemical reactions occurring within the developing embryos and at the same time getting rid off carbon-dioxide.
- ☆ To provide elasticity for easy oviposition and facilitate hatching at the end of embryogenesis.

The insect eggshell or chorion in mulberry silkworm can be discriminated into four spatially differentiated regions *i.e.* the anterior pole region (which has the specialized area for sperm entry), the posterior pole region, the ventral and dorsal regions and lateral flat region (Kawaguchi *et al.*, 1996). The eggshell is very complex, multilayered, multiregional, supra-molecular structure. There are three major layers in any insect shell: the vitelline membrane, the endocrine and the exocrine. The exterior layer is thin with rough web-like sulci and stripes; the second layer is much thicker than exterior with a structure of stripes and granules, the inner layer is very thin with a construction of multipores. Eggshell is made mainly of chorionin, which is composed of scleroprotein and some sacchriode and lipoid. The major amino acids composed of scleroprotein are presented in Table 4.2. The eggshell is produced by mesoderm follicular epithelial cells, which are the parts of follicle. Each follicle is a unit within the paired ovary. In *Bombyx mori,* ovary is polytrophic meroistic type. The number of ovarioles in each ovary is four and looks like a coiled spring of beads containing eggs. Each ovariole contains approximately 50–60 follicles at various stages of oogenesis. Eggshell formation is a gradual process involving certain cellular activities. Such activities have been recognized and analyzed morphologically and biochemically in *Antheraea polyphemus* and *Bombyx mori.*

Table 4.2: Amino Acids Content in the Chorion of *Bombyx mori eggs*

Amino Acid	*Outer Layer*	*Middle Layer*	*Inner Layer*
Glycine	36.8	35.0	18.2
Alanine	8.5	3.1	8.3
Serine	3.2	4.5	6.7
Tyrosine	8.1	1.5	2.2
Valine	6.1	5.1	6.0
Asparagine	4.0	4.9	9.6
Gutamic acid	3.7	3.5	11.6
Threonine	3.3	2.0	5.1
Lysine	0.1	0.4	2.5
Phenylalanine	1.9	1.2	0.15
Isoleucine	3.3	2.1	3.8
Arginine	2.3	3.2	3.5
Leucine	6.4	1.4	7.0
Proline	4.3	2.3	10.1
Histidine	0.4	0.0	1.1
Methionine	0.0	0.4	1.1
Cysteine	5.8	29.6	0.5
Tryptophane	1.8	0.8	0.7

Chorion synthesis starts with the formation of 'trabecular layer' on the outer side of vitelline membrane by the release of micro particles from follicular epithelium. Several layers of helicoidally arranged fibrillar lamellae cover the trabecular layer. The outer region of egg contains unique 'osmiophilic' layer with 15–20 lamellae and very narrow micropyle. Rough surface endoplasmic reticule and Golgi bodies increase greatly in number at the later stage of inner layer formation and are dispersed throughout the cytoplasm of follicle cells. The middle layer of chorion is formed by secretary granules, which are recognized in Golgi regions. The middle layer of chorion consists of numerous fibrous elements that are seemed to be transformed from the secretary granules. Degenerative changes take place remarkably in the components of follicle cells, after the formation of chorion.

Micropyle and Vitelline Membrane

The surface of micropyler region shows a petal like pattern, with the opening of micropyler canals at its centre. The micropyler canals are normally 3–4 in number but in some cases 6–8 canals have also been observed. Each micropyler canal can be divided into two parts–a tube penetrating the chorion and a protrusion connecting the tube to the vitelline membrane. The former is designated as 'ectomicropyle' and the latter 'endomicropyle'. In addition to micropyle, there are about 5000 -10000 tiny pores around the egg surface. These pores permit air entry to meet the requirement of oxygen during embryonic development.

Beneath the chorion is very thin vitelline membrane enclosing the serosa and semi-liquid fluid (yolk), which fills the eggs. The vitelline membrane is secreted uniformly and its formation starts after the degeneration of nurse cells *i.e.* at the time when oocyte occupies almost the entire egg chamber. The vitelline membrane is composed of an electron dense outer layer and an inner layer containing abundant irregularly shaped electron dense granules of various sizes.

Composition of Silkworm Eggs

The biochemical composition of silkworm eggs varies depending on the variety, season and environmental conditions. However, normal composition of eggs is presented below (Table 4.3).

Table 4.3: Biochemical Composition of Silkworm Eggs

Sl.No.	Component	Per cent
1.	Protein	~ 10.00
2.	Lipid	~ 8.50
3.	Glycogen	2.00 ~ 5.00
4.	Chorion	14.00 ~18.00
5.	Water	63.00 ~ 65.00

II. Development of Egg (Embryogenesis)

Different developmental stages of mulberry silkworm eggs (Stage 1–30) are shown in Figure 4.1 and Table 4.4.

Fertilization

The fertilization is a life-saving event, which saves sperm and ovum from certain death, with the union of the two gametes acting as an evolutionary springboard to launch a new individual into the future. When male and female moth of *Bombyx mori* copulates, ejaculation occurs and innumerable spermatozoa enter into the bursa copulatrix of female moth's reproductive organ. Through ductus seminalis, ductus tortuosus and receptaculum seminis, these sperms reaches the lower part of ovipositor and wait for an egg coming down to ovarian tube. Just before the egg is being laid, the spermatozoa enter into it through the micropyle.

Two layers–the vilelline membrane and the chorion, surround the mature egg of the silkworm. The chorion consist two groups of cells–the outer layer and the inner layer. The inner layer and the inner portion of outer layer are mechanically soft and pliable, while the outer portion of outer layer is mechanically rigid. The morphological feature of the chorion that allows penetration of sperm is micropyler apparatus, situated at the anterior end of the egg. Further, the morphological organization of chorionic layer is extremely modified in the region of micropyle. The first maturation (reduction) division begins just before oviposition and terminates 5–6 minutes after the eggs are being laid. Thus, the nuclei in the egg, which have been just oviposited, are in anaphase stage of first reduction division. The female nucleus is divided into two, of which one is being eliminated under the vitelline membrane (the first polar

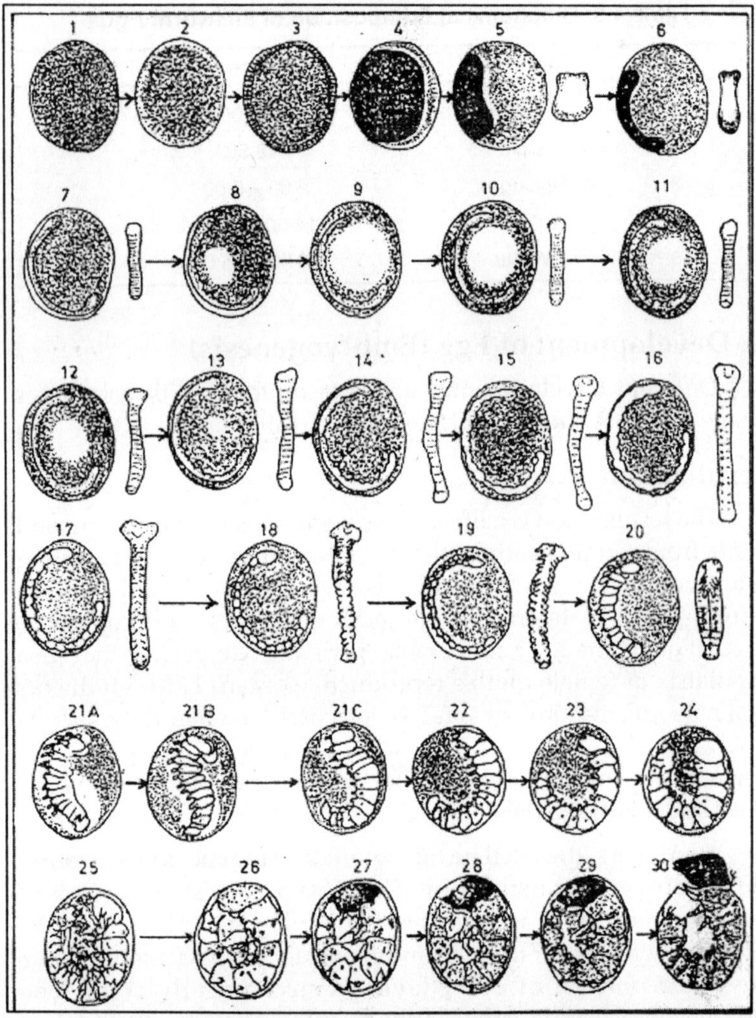

Figure 4.1: Different Stages of Embryonic Development of Silkworm (*Bombyx mori*) Eggs

body). The remaining one again divides into two, one of which is again being eliminated (second polar body). This second maturation division takes place in about 60 minutes after oviposition and is completed in about next 20 minutes, releasing the second polar body. This whole process of reduction division takes approximately 90–100 minutes from oviposition.

Table 4.4: Developmental Events During Preservation of Silkworm Eggs of Bivoltine Breeds

Sl.No.	*Temp. (°C)*	*Stage of Embryo*	*Physio-Morphological Status of Silkworm Eggs*
1.	25	**Pre-diapause stage**	
		Stage 1 (2 hrs)	Fertilization
		Stage 2(10hrs)	Cleavage
		Stage 3 (10–20 hrs)	Cellular blastoderm formation
		Stage 4 (20–24 hrs)	Germ band formation; commonly called Dharuma stage embryo. The margins of the germ band are cut off from the extra embryonic area and the cells expand over the germ band to form serosa. Optimum age for HCl treatment; maximum oxygen consumption.
		Stage 5 (24–30 hrs)	Cephalic lobe and caudal lobe formation; primitive groove appearance; pigmentation in serosal cells.
		Stage 6 (30–40 hrs)	Spoon shaped embryo; formation of ectoderm, mesoderm and segmentation of mesoderm.
		Stage 7 (40–72 hrs) (Chilling Stage)	Appearance of 18 segments; respiration switch over from aerobic to anaerobic; pentose phosphate pathway with glycolysis; operates to convert glycogen into polyols; steep decline in oxygen consumption.
		Diapause stage	
		Stage 8 (14 days) (Diapause Stage-1)	Yolk cell migration; embryo surrounded by yolk cells; mitotic division stops on 4th day; continued steep decline in glycogen content.

Contd...

Table 4.4–Contd...

Sl.No.	*Temp. (°C)*	*Stage of Embryo*	*Physio-Morphological Status of Silkworm Eggs*
		Stage 9 (60 day) (Diapause Stage 2)	Wider central region lacking yolk cells; rate of respiration comes down to 2–5 per cent, arrest of water loss.
2.	20, 15, 10	Stage 9 (continues)	Total arrest of water loss; eggs are under anaerobic respiration; rate of metabolism is diminished. Eggs do hatch unless they are transferred at 5°C.
3.	5	Stage 10 and 11 (Hibernating Stage)	Cryoprotection of embryo by sorbitol and glycerol; 60 days onwards starts declining in the levels of sorbitol by the activation of NAD-sorbitol dehydrogenase for the re-synthesis of glycogen (a sign of termination of diapause); starts oxygen consumption (aerobic respiration).
4.	2.5, 1.0, 0, –2.5	Stage 11 (continues)	Low rate of metabolism to save energy; no embryonic growth. It means releasing of eggs for synchronization and incubation.
5.	–10, –15, –20	Stage 11 (continues)	Super cooling of eggs, silkworm eggs can survive by avoiding freezing and adopting super cooling.
6.	15	**Intermediate care** (Stage 11, 12 and 13)	Post diapause development; aerobic respiration increases, embryonic growth by elongation of embryo and not by increasing the number of cells by mitotic division. Mesodermal segments visible, tails differentiation starts.
		Stage 14	Critical stage I. Embryo is longer and slender. Tail region differentiated. Commonly called as Hei-A stage.18 mesodermal bands can be easily observed.
		Stage 15 (3–4 days)	Critical stage II (fit for double refrigeration).

Contd...

Table 4.4–Contd...

Sl.No.	Temp. (°C)	Stage of Embryo	Physio-Morphological Status of Silkworm Eggs
7.	2.5	Stage 15 (double refrigeration) (30–40 days)	The embryo is in its longest stage called longest embryo or Hei-B stage. It occupies 3/4 circumference of the egg. Head pigmentation is clearly visible. 18 mesodermal segments are clearly visible.
8.	25	**Incubation Stage** (Organogenesis) (Stage 16)	16 hrs light & 8 hrs dark/day to maintain bivoltine characters; starts losing 1–2 per cent of water per day. Hence maintain 75 ± 5 per cent RH to avoid desiccation; within 0.5 hr after shifting from 15°C to 25°C the neural groove appears; starts gradual increase in oxygen consumption. This stage is commonly called incubation stage.
9.	25	Stage 17 (1.5–2 days)	Appendages appear in the form of buds in the protocephalon, gnathal and thoracic region. Segmentation of the body region is clearly visible. From this stage onwards 400–500 lux light with 100 cm distance is to be provided.
		Stage 18 (2.5–3 days)	Mandible, maxillary, labial and three thoracic segments become clear. Stomodaeam is clearly visible. 4 pairs of abdominal buds are seen from 3–6 abdominal segments.
		Stage 19	The labial and antennal rudiments appear as paired hemispherical lobes. Thoracic legs show segmentation. Shortening of embryo.
10.	25	Stage 20 (3–4 days)	Further development of appendages, head and thorax distinguishable, embryo is further shortened, better to transport the eggs before attaining the blastokinesis stage.
		Stage 21 (A, B and C) (4.5–5 days)	Morphogenetic movement of embryo, early, middle and late blastokinesis, completion of dorsal closure.

Contd...

Table 4.4–Contd...

Sl.No.	Temp. (°C)	Stage of Embryo	Physio-Morphological Status of Silkworm Eggs
		Stage 22 (5 days)	Complete embryonic reversal (blastokinesis completed). Morphogenetic movement of the cephalognathal region begins to form head.
		Stage 23 (6 days)	Thoracic and abdominal appendages clearly visible. Steep rise in oxygen consumption and CO_2 release.
		Stage 24	Tubercles appear on the body of the embryo. Setae formation. Abdominal legs clawed, caudal horns also develop.
		Stage 25	Formation of taenidia inside the tracheae and tracheal, initiation of sperm tube formation.
		Stage 26 (8 days) (head pigmentation–I)	Up to 0.5 per cent CO_2 in air tolerable, beyond this causes uneven development, irregular and delayed hatching, head capsule darkening.
		Stage 27 (head pigmentation–II)	Sensitive to high CO_2 level and low humidity. Most ideal stage for black boxing.
11.	25	Stage 28 (9 days) (early body pigmentation)	Darkening of thorax and abdomen, dark coloration helps in obtaining the synchronized hatching stage.
		Stage 29 (10 days) (late body pigmentation)	More darkened body, young larva completed, Sensitive to low RH and high CO_2 level, hence eggs may die in this stage.
		Stage 30 (10 -11 days)	Hatching of young larvae within 2–3 hrs after light exposure.

As soon as reduction division is completed only one nucleus will be seen in the egg, which becomes a female pronucleus. During the interval when all these changes take place in egg nucleus, the sperm sheds its tail. The central body of sperm divides into two and the anterior part swells into spherical body called male pronucleus. The female and male pronuclei unite to form the zygote. This process is known as syngamy, which is being completed within 2 hrs of deposition of egg (Stage 1). Thus after fertilization, the zygote restores the original chromosome number of the organism (2n = 56). Though, the number of spermatozoa entering the egg may be several, only one of them penetrates into an ovum and fertilizes the female pronucleus and the rest degenerate and are absorbed.

Cleavage and Blastoderm Formation

As in most insects, the silkworm egg also undergoes superficial cleavage. The content of eggs consists of two parts–the protoplasm and the endoplasm. Protoplasm forms a reticulum that pervades the substance of egg and forms a bounding layer beneath the vitelline membrane. This is termed as periplasm, which is free from yolk spheres. The endoplasm (yolk), which occupies the central part of the egg, is contained within the protoplasmic reticulum. The zygote nucleus forms many nuclei by synchronous mitotic divisions, which is being reported 10 times approximately within 10 hrs of oviposition (Stage 2). Afterwards, luminous circle of microtubules appear around the cleavage nuclei. Takesue *et al.* (1980) suggested that these microtubules are responsible for the movement of cleavage nuclei towards the egg surface.

These dividing nuclei surrounded by protoplasm are transformed into cells. Subsequently, greater number of these cleavage cells migrates towards the periphery of the egg, merge in the periplasm and form a layer directly beneath the vitelline membrane called blastoderm. Some of the cleavage nuclei do not move to the periphery but persists and form yolk cells or vitellophages. These vitellophages have the function of supplying nutrients to the embryo. The nuclei projecting into the periplasm are then pinched off from the yolk system and approximately 10–20 hrs of oviposition, uniform blastoderm appears (Stage 3).

Germ Band and Amnion Formation

The cells, which form the blastoderm, are alike. About 20 hrs of oviposition, part of blastodermal cell divides asynchronously several times and become thicker and cuboidal in appearance on the ventral side. These cells changes gradually from cuboidal to cylindrical shape and then become distinguishable from the other part called germ band or germinal cord (Stage 4). The yolk membrane forms enfolding during the late germ band stage. Afterwards, yolk system is divided into many masses, each enclosing one or several nuclei and yolk organelles. This process is known as yolk segmentation.

Parallel to the major axis of the germ band, envelope like structure appears from both side and partly covers the germ band. These are amniotic folds. Amniotic fold consists of two layers (inner and outer), grows inwards, towards each other, and finally reaches to the centre of the germ band. The outer layer of the fold extends from all the sides and meets to form the serosa, which covers entire surface of the egg. The inner layer of the folds meets and unites in the thick portion of the blastoderm to form amnion. The cavity formed by the two amniotic folds in which germ band lies are known as amniotic cavity. The thick portion of the blastoderm covered by amnion develops into embryo.

Organogenesis

A long and narrow depression (primitive groove) in the middle on the ventral side of the germ band is formed (Stage 5). The primitive groove invaginates deeper and deeper and the tips of invaginations are brought together covering the groove. After completion of covering, the groove cannot be seen externally except leaving a small aperture (blastopore) on the head. Beneath this primitive groove, a group of spherical cells are found which leads to the formation of mesoderm or endomesoderm. The ventral plate, which is an elliptic disc at the end of 24 hrs of oviposition, changes into spoon-shape at the end of 35 hrs and become enlarged (Stage 6). Afterwards, the germ band sinks into yolk and by divisions of yolk cells, primary yolk cells are formed which later leads to the formation of secondary yolk cells. The anterior end of spoon (swollen part at the head region) is called 'procephalic' or 'protocephalic' lobe and that at the tail end is 'caudal lobe'. At this stage, mesoderm becomes morphologically distinguishable from the ectoderm.

At the end of approximately 2 days of oviposition, neural groove appears in embryos (Stage 16, non-diapause}. Appendages development in gnathal and thoracic regions starts at the very next stage (Stage 17). After three days of oviposition, abdominal limb buds are formed from third to sixth abdominal segments (Stage 18). Paired gonad analages also appears at the same time. Next stage of development indicates beginning of embryo shortening (Stage 19), invagination of spiracles and silk glands besides formation of strand like gonad anlages. After four days of oviposition, the head and thorax becomes differentiated (Stage 20), with the grouping of gnathal appendages.

The anterior six segments of embryo fuse together to form the head. During fusion, two segments disappears and thus leaving four in the head region. Thoracic appendages develop into three segmented legs. In the abdominal region, except the five pairs of appendages, which persists later in the larva, remaining disappears during the course of development. Last three segments of the abdomen fuse together to form single segment and thus leaving abdomen with only nine segments. Blastokinesis begins at this stage. Actually, blastokinesis is the characteristic of all those insects whose eggs are rich in yolk and the germ band is invaginated therein. Blastokinesis occurs 4–5 days after oviposition. The embryo starts to move around as it seeks its correct position in the egg (Stage 21A, B & C). For instance the embryo, which was in the supine position with its abdomen facing outer of the egg turns round, so that its abdomen faces the inner side. This whole process is called blastokinesis. The anterior and posterior ectodermal invaginations (stomodaeum and proctodaeum) forms the fore intestine and hind intestine respectively. During this stage, the embryo, located on the ventral side of the egg, moves towards the dorsal side.

After blastokinesis, histogenesis proceeds actively and continues till the larva hatches. The first embryonic molt takes place just after blastokinesis followed by second molt immediately, after which the larval cuticle is secreted (Stage 24). After revolution, various organs are gradually formed and eggs hatch in about 5 days later. The number of embryonic segments in *B. mori* is reported to be 18. Of these, four anterior segments form the head, three segments form the thorax and the other eleven segments form the abdomen.

Mesoderm cells, forms four pairs of appendages on the head region and three pairs on the thoracic region. Besides these seven pairs of appendages, the mesodermic cells also form one pair of appendage on each of the eleven abdominal segments. Of these eleven pairs of appendages on the abdominal region, only 3rd, 4th, 5th, 6th and 11th abdominal segments appendages develop further and remaining six pairs disappear during the course of development (Stage 18).

While antenna, skin, mouth parts, thoracic leg, abdominal leg, caudal leg, silk gland and other dermo-glands besides fore and hind intestine, nervous system, trachea, sense organs, malphigian tube and external genetalia are formed from ectodermal cells, internal genetalia, muscles, dorsal vessels, sub-esophageal gland, corpuscles, pericardial cells, fat tissues originate from mesoderm. In diapausing embryos, cell division is arrested and embryogenesis ceases immediately after formation of the cephalic lobe and segmentation of mesoderm (Stage 7). Yolk cells and yolk granules change their shape and properties as a part of diapause process. The most striking feature of the diapausing eggs is the dark colouration of serosal cells due to the formation of ommochrome pigment. When diapause is terminated (Stage 15), embryogenesis resumes and follows the pattern as in non-diapause eggs. At stage 6, the development rate in diapausing eggs becomes slower than in non-diapausing eggs. Hence, stage 8–15 is absent in non-diapausing eggs and all other stages of development in both the cases are similar. Further, in diapausing eggs, stage of development from 16 onwards is delayed approximately one day in comparison to non-diapausing eggs.

Respiration through Eggshell

Respiration is a fundamental biological phenomenon necessary for the maintenance of biochemical reactions, which are the characteristic of the living systems. The gas exchange process is passive and is based upon diffusion governed by coefficient and relative partial pressure of oxygen on either side of the eggshell. Respiratory canals are funnel shaped tubes, which becomes narrower as it enters inward. These respiratory canals are aeropyles and are responsible for gas exchange process during embryogenesis. These aeropyles are tiny holes transversing relatively thick chorion and conducting the ambient air directly to the trabecular layer and from there to the oocyte. These aeropyles functions during embryogenesis in gas exchange as per 'Bernoulli Velocity Gradients Theory'. The aeropyles are distributed in large numbers over the entire surface of the egg except around the micropyle.

Structurally, aeropyle openings are somehow elevated or surrounded by crowns. These respiratory devices carry the ambient air to the internal trabecular layer for both storage and immediate consumption. Oxygen moves within the trabecular layer and cross the inner layers (endochorion and vitelline membrane) and reaches to the developing oocyte. The inner membranes are structurally porous, so that the oxygen can pass through, where the metabolic processes are taking place. Moreover, if the ambient environment is poor in oxygen, having a partial pressure lower than that within the embryo; it may prove fatal for the development. Carbon dioxide is driven easily from inside, since its partial pressure in the environment is virtually zero.

Since, the oxygen and carbon dioxide molecules are bigger than the water molecules; it was not possible for biological systems to evolve a kind of membrane, which would be permeable only to oxygen and carbon dioxide, but not to water. Hence, the loss in water content during embryogenesis cannot be ruled out. Number of workers has described the phenomenon of gradual loss in weight of fertilized eggs during the course of their embryonic development. Therefore, the weight of silkworm eggs is not a constant factor during the course of development. The maximum loss in weight of eggs has been reported on the day of hatching. The loss in weight of eggs in the course of embryogenesis is partly due to the exhalation of carbon dioxide and partly due to the loss in moisture content (Mathur and Singh, 1990). The difference in egg weight is also due to energy loss during embryogenesis.

Embryo Isolation

Embryological studies in sericulturally advanced countries are well practiced. It is a technology aimed in solving practical problems related to identification of suitable embryonic stages in order to develop an effective and efficient model for long-term preservation of silkworm eggs. The eggs are to be handled in accordance to their embryonic stages to maintain or retain the potency or viability, as each stage has its own tolerance limits for temperature and humidity conditions. Embryo isolation technique is very crucial in monitoring and identifying the different stages of development of eggs during aestivation, hibernation, preservation and release from cold storages or re-refrigeration during long-term preservation and incubation.

Advantages of Embryo Isolation

- Helps in long-term preservation of bivoltine as well as multivoltine eggs.
- Helps as an appropriate indicator of time for release of hibernated eggs.
- Helps in handling of eggs in accordance with the need of environmental conditions.
- Helps to forecast the exact date and hatching percentage well in advance.
- Helps in identifying the specific stages which are sensitive or resistant to various egg handling techniques.
- Helps to judge whether eggs are properly handled or not.

Techniques of Embryo Isolation

Different techniques are followed in isolation of embryos for identification of various developmental stages. The techniques are:

(*a*) Hot Water Fixation Technique

- Treat the silkworm eggs in hot water.
- Dry the silkworm eggs under fan.
- Cut the egg shell from the micropyler region laterally using a sharp blade under a dissection stereomicroscope.
- Squirt water with force using Pasteur pipette on cut eggs placed in Petri dish with water to release the embryos.
- Transfer the released embryos carefully onto a micro-slide.
- Remove the yolk particles if any sticking to the embryos using the soft brush.
- Observe under microscope to identify the stages of embryonic development.
- Permanent slides can be prepared using Borax carmine or Haematoxylin stain.
- In case of loose eggs, the eggs are to be fixed in paraffin wax in Petri dish and cut.

(*b*) Potassium Hydroxide Technique (KOH)

- Prepare 2–4 per cent KOH solution.
- Boil the KOH solution up to boiling point (60°C).

Table 4.5: Hot Water Fixation Technique for Embryo Isolation

Sl.No.	*Embryonic Developmental Stage*	*Temperature of Water (°C)*	*Dipping Duration (Minutes)*
1.	24 hrs (1st day) (Stage 5)	70	3–4
2.	36 hrs (2nd day up to blastokinesis) (Stage 18)	75	4–5
3.	Blastokinesis to head pigmentation (Stage 22–27)	76	5
4.	Head pigmentation to blue egg stage (Stage 27–29)	77	6

- ☆ Dip the egg sheet or loose eggs in boiling KOH ((60°C) solution for 30 seconds.
- ☆ Care is to be taken that the embryos do not get dissolved in KOH solution.
- ☆ Transfer the content under tap water.
- ☆ Squirt water with force using dropper over eggs to release embryos.
- ☆ Transfer the released embryos carefully on to a micro-glass slide with the help of a dropper.
- ☆ Observe under microscope.
- ☆ Identify the stage of embryonic development.

(*c*) Transparent Technique

- ☆ Whole eggs are fixed in carmine for 30 minutes at 30°C.
- ☆ Transfer and preserve eggs then in 70 per cent alcohol.
- ☆ De-chorinised the eggs.
- ☆ Transfer the egg in 1 per cent HCl for 12 hrs at 50°C.
- ☆ Embryo becomes visible.
- ☆ Stain with Borax carmine or Haematoxylin.

Permanent Slide Preparation

Permanent slides for various developmental stages can be prepared for future reference. The embryos are isolated from egg shell adopting either hot water or KoH embryo isolation technique. The isolated embryos free from yolk are washed in distilled water.

The washed embryos are passed through different grades of alcohol as indicated below (Table 4.6).

Table 4.6: Dehydration of Embryos for Staining

Sl.No.	Alcohol (Per cent)	Dipping Duration (Minutes)
1.	30	20
2.	50	10
3.	70	10

After washing the embryos in 70 per cent alcohol for 10 minutes, stain in Borax carmine for 5–20 minutes keeping in view of the intensity of colour required. After required colour is developed, embryos are washed again in 70 per cent alcohol for 2–3 minutes. The stained embryos are again passed through different grades of alcohol as indicated below (Table 4.7).

Table 4.7: Dehydration of Embryos for Mounting

Sl.No.	Alcohol (Per cent)	Dipping Duration (Minutes)
1.	80	10
2.	90	10
3.	Absolute	2-10

The embryos are again washed in absolute alcohol for 2–3 minutes and then passed through xylene for 1–2 minutes. The embryos are then mounted with DPX and covered with cover-slip avoiding air bubbles to enter the slide. The slide thus prepared is kept in an oven for 2–3 days maintained at 40°C.

Suitable Embryonic Stages for Cold Preservation and Release

- ☆ Embryonic stages ideal for chilling is Stage 5 (24 hrs) to Stage 7 (48 hrs).
- ☆ Embryonic stage ideal for different hibernation schedule is Stage 8 (diapause stage).
- ☆ During cold storage, when diapause is in the stage of termination, the embryo reaches 10/11 stage depending

on the duration of cold storage. When reaches Stage 11, it is fit for release and incubation.

☆ Embryonic developmental stages decide when long-term preservation of bivoltine eggs and double refrigeration can be undertaken. For this, when embryo reaches Stage 11 are released from cold storages and transferred at intermediate temperature (15°C) for 3 days. During this period, the embryos reach Stage 15 which is suitable for re-refrigeration at 2.5°C temperature for 10–40 days. The technique for prolonging the refrigeration is known as intermediate care. During intermediate care, if embryos reach Stage 16, instead of Stage 15, then they are not suitable for re-refrigeration or double refrigeration and such eggs can be incubated, brushed and reared. By adopting intermediate care, embryos retain its vitality and withstand long-term cold storage besides ensuring hatching in a single day. Embryonic Stage 5 at 10°C and Stage 15 at 5°C are suitable for long-term preservation of cross breed multivoltine eggs. However, cold storage of multivoltine eggs should be avoided as-far-as possible.

Chapter 5

DIAPAUSE AND BIOCHEMICAL MODULATIONS

The nature of diapause in mulberry silkworm, *Bombyx mori* is primarily determined by genetic characters and endocrinological mechanisms, mediated by environmental factors such as temperature, humidity and photoperiod and is almost maternal. Hibernating potency besides nucleic acids and carbohydrate metabolism is also equally responsible for induction, initiation, maintenance and termination of diapause. Since mulberry silkworm enter diapause as embryo, the duration of egg life depends on the duration of embryonic diapause which, represents a syndrome of physiological and biochemical events.

Embryonic diapause in *Bombyx mori* induced by active secretion of sub-esophageal ganglion is attributed to the metabolic adjustment, which serves to bring about a new physiological state. Metabolic conversion of trehalose to glycogen at the induction, glycogen to sorbitol at initiation and sorbitol to glycogen at termination of diapause is correlated and in each metabolic shift, a key enzyme becomes active in response to hormonal and environmental stimulation.

Colouration or pigmentation of insects is mainly due to melanin, pterine, pigments, ommochrome and carotenoid. In some insects, colouration is closely correlated with diapause and colour pattern is sometimes used as an index to distinguish diapause from non-diapause. Among these pigments, ommochrome is well documented in relation to embryonic diapause. Diapause development is under the control of endocrinological mechanism mediated by environmental conditions such as temperature, photoperiod, humidity etc. Hence, diapause is characterized by an active mechanism for adaptation to the adverse seasons.

I. Diapause

Diapause can occur at embryonic *(Bombyx mori, Aedes aegypti),* larval *(Cydia pomonella, Gilpinia polytoma),* pupal *(Antheraea mylitta, Hyalophora cecropia)* or adult *(Eurydema* sp., *Leptinotarsa decemlineata)* reproductive stages of life cycle and the stages are characteristically fixed in each species (Table 5.1).

Table 5.1: Diapause Nature in Different Insect Species (*c.f.* Singh and Saratchandra, 2002)

Nature of Diapause	*Example*
Embryonic diapause	*Bombyx mori, Melanoplus differntialis, Aedes aegypti, Aeschna mixta.*
Larval diapause	*Cydia pomonella, Anax* sp., *Epistrophe bifasiata, Gilpinia polytoma, Cephus cinctus, Lucilia Caesar.*
Pupal diapause	*Antheraea pemyi, Hyalophora cecropia, Antheraea mylitta, Philosamia cynthia, Saturnia pavonia, Mimas tiliae.*
Adult diapause	*Eurydema* sp., *Pterostichus* sp., *Leptinotarsa decemlineata, Musca autumnalis.*

Nature of Diapause

Genetic characters primarily determine the nature of diapause in the silkworm *(Bombyx mori).* The insects having only one generation in a year (univoltine), embryonic diapause occurs in all generations regardless of environmental conditions, is known as 'obligatory diapause', while those having two or more generations in a year (bivoltine or polyvoltine), diapause is expressed in next generation when the mother receives specific stimuli from environment is termed 'facultative diapause'. The univoltine forms lay only hibernating (diapausing) eggs, which are also called 'Kurodane eggs', and multivoltine forms lay only non-hibernating eggs which are known as 'Nemadane eggs', while the behaviour of bivoltine are intermediate. Bivoltine silkworm breeds lay non-hibernating eggs during first generation and hibernating eggs in the next generation, which hatches out in the following spring and thus producing only two generations in a year. Practically, univoltine and bivoltine silkworm breeds produce superior cocoons to those of multivoltines in both quality and quantity. Generally, silkworms programmed to lay diapause eggs are superior to those that of non-diapause eggs. The silkworm eggs of diapauses type are pigmented (dark brown) due to presence of ommochrome formed in serosa cells, while non-diapause eggs are light yellow / white due to lack of this pigment. Tryptophane metabolism in insects has been studied extensively and the pathway from tryptophane to ommochrome has been reviewed in detail by Linzen (1974). In general, three major metabolites directly involved in the biosynthesis of the pigments are formyl-kynurenine, kynurenine and 3-hydroxy kynurenine. The corresponding enzymes responsible for their production are kynurenine-formamidase, kynureninase and kynurenine-3 hydroxylase. Ommochrome is synthesized in developing ovaries and other tissues such as fat body during tryptophane metabolism as shown below:

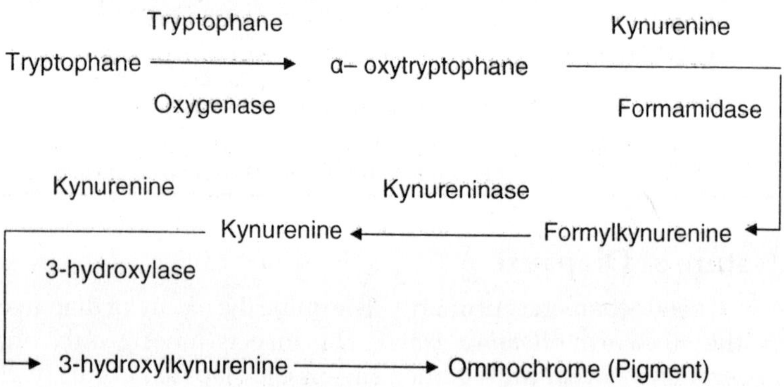

Genetical Studies on Diapause

The application of genetics to diapause phenomenon was considered in *B. mori* during early Mendelian era, and since then, it has been studied by many workers. Nagatomo assumed a series of three sex-linked alleles, Hs, Hs2 and hs and three pairs of autosomal genes H1h1, H2h2 and H3h3 responsible for the hibernation and reported different kinds of voltinism in relation to 'Hibernating potency value' (H.V.) (Table 5.2). The eggs with 0–1 HV value always lay non-hibernating eggs (multivoltine), while those with 11–12 HV value will lay only hibernating eggs (univoltine). Bivoltine silkworm breeds have HV value in between 5–7. The silkworm eggs with HV value of 2–4 are intermediate between multivoltine and bivoltine and lays some non-hibernating eggs at high incubation temperature. Contrary to this, the silkworm eggs with HV value of 8–10 are intermediate between bivoltine and univoltine and lay both hibernating and non-hibernating eggs at low temperature incubation. Morohoshi and Mezaki (1957), on the other hand believe, the major genes determining voltinism to be a series of three multiple alleles (V^1 $+^V$ and V^3) although their action is affected to a great extent by sex-linked maturity genes (Lm–late maturity). Morohoshi (1976) proposed a hypothetical genetic balance theory between major genes and constant gene of Lm to Lm^e (early maturity) as the basis of voltinism phenomenon. Genetical experiments have shown the major factor controlling diapause belongs to two loci *i.e. Lm* and *V*, and these loci as well as other loci control process other than synthesis of DH. Considering the various investigations on the genetics of voltinism, carried out by many workers, Subramanya and Murakami (1994) stated that voltinism is maternally inherited phenomenon controlled by sex-linked genes regardless of artificial treatment of eggs.

Table 5.2: Relationship Between Hibernating Potency Value and Voltinism in Silkworm (*Bombyx mori*)

Hibernating Potency Value	*Voltinism*	*Characteristics*
0–1	Multivoltine	Always lay non-hibernating eggs.
2–4	Intermediate between multi and bivoltine	Lay some non-hibernating eggs at high incubation temperature.
5–7	Bivoltine	Lay non-hibernating eggs at low temperature and hibernating eggs at high incubation temperature.
8–10	Intermediate between bivoltine and univoltine	Lay both hibernating and non-hibernating eggs at low incubation temperature.
11–12	Univoltine	Always lay hibernating eggs.

Source: Nagatomo, 1953.

Factors Affecting Diapause Nature

The silkworm *(Bombyx mori)* diapause determination is almost maternal, that is, temperature and photoperiod are most efficient at the embryonic stage of previous generation (Table 5.3) and only supplemental in the post-embryonic stages. The sensitive embryonic stages begin just after blastokinesis. Incubation of eggs of bivoltine breeds at high temperature results in the induction of diapause eggs in the next generation and a low temperature incubation results in the production of non-diapause eggs. Incubation as low as 15°C causes production of non-diapause eggs in the next generation, whereas, diapause eggs are induced by incubation at 25°C. When eggs are incubated at intermediate temperature of 20°C, the developmental fate remains undetermined in the embryos. High temperature at younger larval stages and low temperature at late larval stages acts to induce diapause eggs (Tazima, 1978). Photoperiod is one of the principle physical factors regulating diapause. In the silkworm, egg diapause is regulated by photoperiod as well as temperature during embryonic stage of the female and is

Table 5.3: Effects of Temperature and Photoperiod on Egg Diapause in *Bombyx mori* (Bivoltine Breeds)

Temperature and Photoperiod during Incubation		*Temperature °C Developmental Stage*		*Resulting Moths Laying Diapause and Non-diapause Eggs*
		I – II	*IV–pupal*	
25°C	Light	25 or 20	25	Diapause
	Dark	25 or 20	25	Diapause
20°C	Light	25 or 20	25 or 20	Diapause
	Dark	20	25	Diapause << Nondiapause
		25	25	Diapause < Nondiapause
		25	20	Diapause >> Nondiapause
15°C	Light	20	25	Diapause < Nondiapause
		25	25	Diapause > Nondiapause
		25	20	Diapause >> Nondiapause
	Dark	25 or 20	25 or 20	Nondiapause

Note: Light : 16 hours light and 8 hours' dark; Dark: 16 hours dark and 8 hours light.

completely independent of photoperiod during post-embryonic development. Thus, photoperiod becomes effective in regulation to development only when eggs are incubated at an intermediate temperature. In these eggs, long photoperiod causes induction of diapause and short photoperiod non-arrested state of development (Table 5.3). In silkworm, injection of Uranyl nitrate, Quabain and 5'-AMP into females destined to lay diapause eggs caused them to lay non-diapausing eggs. Converse was demonstrated by injection of KCl and Quabain into pupae of non-diapause type. However, despite extensive studies, the mechanism of action of these chemicals remains unknown.

Diapause Induction

The diapause determination is almost maternal in mulberry silkworm *(Bombyx mori), i.e.* temperature and photoperiod are most efficient at the embryonic stage of previous generation (Table 5.4) and only supplemental in the post-embryonic stages. The sensitive embryonic stages begin just after blastokinesis. Incubation of bivoltine eggs at high temperature results in induction of diapause in the next generation and low temperature incubation results in the production of non-diapause eggs (*c.f.*, Yokoyama, 1973). Incubation as low as 15°C causes production of non-diapause eggs in the next generation, whereas, diapause eggs are induced by incubation at 25°C. When eggs are incubated at intermediate temperature of 20°C, the developmental fate remains undetermined in the embryos. High temperature at younger larval stages and low temperature at late larval stages acts to induce diapause eggs (Tazima, 1978). Egg diapause is regulated by photoperiod as well as temperature during embryonic stage of the female and is completely independent of photoperiod during post-embryonic development. Thus, photoperiod becomes effective in regulation to development only when eggs are incubated at an intermediate temperature. In these eggs, long photoperiod causes induction of diapause and short photoperiod non-arrested state of development.

In silkworm, injection of Uranyl nitrate (Hasegawa, 1957), quabain (Takeda and Hasegawa, 1975) and 5'-AMP (Suzuki *et al.*, 1978) into females destined to lay diapause eggs caused them to lay non-diapausing eggs. Converse was demonstrated by injection of KCI and quabain into pupae of non-diapause type (Takeda and Hasegawa, 1976). However, despite extensive studies, the

Table 5.4: Relationship Between the Conditions of Incubation, Rearing, Mounting and Diapausing Character of Resultant Egg in Bivoltine Silkworm Races (*c.f.*, Tazima, 1978)

Developmental Stage	*Environmental Conditions*	*Effect*	*Remarks*
During incubation	**Temperature**		
	High	Diapausing	Whole eggs at 25°C or above
	Low	Non-diapausing	Whole eggs at 15°C or above
	Humidity		
	Humid/Dry	Diapausing Non-diapausing	Humidity affects voltinism only when the incubation temperature is at 15-25°C.
	Light	Diapausing	Light (18 hrs or more a day). Photoperiod affects only when incubated at 15-25°C
	Dark	Non-diapausing	In case of 12 hrs darkness a day
Early larva	**Temperature**		
	High	Diapausing	Only when incubated
	Low	Non-diapausing	at 15-25°C
Late larva	**Temperature**		
	High	Non-diapausing	Only when incubated
	Low	Diapausing	at 15-25°C
Mounting and later	**Temperature**		
	High	Non-diapausing/	Only when incubated
	Low	Diapausing	at 15-25°C

mechanism of action of these chemicals remains unknown. Ando (1974) classified diapause into three groups based on the requirement of water for embryonic development *viz.*, sufficient water is stored in the egg at the time of oviposition, so that embryogenesis is completed till hatching as in *Bombyx mori*; water is absorbed into egg before diapause establishment and utilized for post-diapause development as in crickets; and water is absorbed mainly after diapause termination for the completion of embryogenesis as in grasshoppers. In silkworm, *Bombyx mori*, hydrocarbons comprise the major lipid of the egg shell and are believed to take part in water evaporation.

Insect diapause results due to the exposure of sensitive stages to distinct stimuli. This stage is usually fixed in the life history of insect and is characteristics of each species. The time elapsing from diapause induction to its onset may be as long as a complete developmental cycle *i.e.* egg to egg.

Initiation and Establishment of Diapause

In diapause eggs, the oxygen uptake is reported to be 30 µl/g of eggs/hr. Within 4 hrs of oviposition reached to 100 µl/g of eggs/hr to 24 hrs and thereafter declined rapidly reaching to 10 µl/g of eggs/hr on 10th day and 8 µl/g of eggs/hr on 70th day. The decrease in oxygen uptake by day 10 at 25°C indicates the establishment of a stable physiological status of diapause. However, after chilling of eggs for 70 days and releasing at 25°C, increases in oxygen uptake. But when eggs are chilled at 48 hrs after oviposition and released to 25°C, 18 µl/g of eggs/hr is reported until chilling for 50 days. This clearly indicates that decrease in oxygen uptake proceeds even at 5°C and physiological status of eggs exposed to 5°C after incubation at 25°C for 2 days after oviposition is different than in eggs exposed to 5°C for 15 days of oviposition. The ability to break diapause is correlated with the increased rate of oxygen uptake and it is stated that when eggs showed oxygen uptake of 70 µl/g of eggs/hr are incubated at 25°C, more than 80 per cent of larvae hatched. Oxygen uptake will be about 70 µl/g of eggs/hr in acid treated eggs (after one hour) chilled for 30 days at 48 hrs of age of eggs. This value is almost three folds higher found in the eggs before acid treatment (20 µl/g of eggs/hr). At 48 hr of age, diapause is initiated but an intense status of diapause is not established. Initiation of diapause occurs around one to two days after oviposition and a complete physiological status of diapause is established after incubation at 25°C for 10 days after oviposition when the oxygen consumption is at decreasing trend. Also, sorbitol accumulates two days after oviposition and mitotic figures become undetectable in embryonic cells after three to four days of oviposition and this stage is termed as 'pre-diapause'. The oxygen uptake increases to a high level on prolonged chilling which coincides with complete recovery of glycogen from sorbitol and the stage of decreasing glycerol. (Yaginuma and Yamashita, 1977; Yaginuma *et al.*, 1990 a).

Diapause Termination

Exposure of silkworm diapause eggs to oxygen is one of the artificial methods of termination of diapause. The exposure of diapause-destined eggs to oxygen gas can prevent the expression of diapause and eggs resumed embryonic development while under an oxygen-deficient environment non-diapause eggs ceased their development. Oxygen is also indispensable for diapause development during chilling of diapause eggs (Ando, 1974). However, mechanism involved in terminating the egg diapause by exposure to oxygen is still not known. Diapause in silkworm eggs can be terminated to obtain effective hatchability by means of:

I. Cold storage–chilling (hibernation schedule)

II. Hydrochlorisation (hot or cold acid treatment)

III. Cold storage and hydrochlorisation (chilling and treatment).

The application of these methods depends on the programme of hatching desired. Chilling is one of the effective methods of terminating, diapause. Exposure of diapausing silkworm eggs to a temperature as low as 5°C over 60 days completely terminates diapause and embryogenesis resumes when these eggs are transferred at 25°C. Optimum temperature to break the embryonic diapause is in between 5 to 7.5°C. The required chilling duration to break the diapause depends on the time gap the eggs have been kept for aestivation at 25°C after oviposition. Various comprehensive hibernation schedules for preservation of bivoltine eggs for different durations to get the desired hatching at appropriate time have already been recommended and some of them are in-vogue to meet the demand of seed supply throughout the year.

HCl treatment (hydrochlorisation) of silkworm diapause eggs has been the method of choice for blocking the diapause both at commercial and basic research laboratory level. There are two methods of HCl treatment. In the first method 20–24 hrs old oviposited eggs are soaked in HCl solution (specific gravity: 1.075 at 15°C) at 46.1°C for 5 minutes. This avoids the eggs to enter into diapause and hence when incubated at 25°C, larvae hatches in 10 -11 days after treatment. In the second method 20–24 hrs old oviposited eggs are soaked in HCl solution (specific gravity 1.10 at 10°C) at room temperature for 60 to 90 minutes. Sometimes, to block the diapause for longer period in order to get hatching at desired time, chilling followed by acid treatment is also used. In this method, 48 hrs old eggs are first chilled at 5°C for more than 30 days and then soaked in HCl solution (specific gravity 1.10 at 15°C) at 48°C for 5 minutes. This treatment causes diapause eggs to hatch within two months after oviposition.

However, it is still an open question how hydrochlorisation blocks diapause. Park and Yoshitake (1970) suggested that HCl infiltrated into eggs impedes the embryonic protein synthesis system in yolk cells, which results in resumption of embryogenesis. Further, HCl treatment is reported to stimulate the activity of specific esterase isozyme and RNA synthetic activity in treated diapause eggs. Activity of 'esterase A' increases dramatically within 30 minutes of HCl treatment. The esterase attains maximum activity before the eggs are competent to develop, and thus 'esterase A' is pre-requisite for resumption of development.

Diapause Control–Hormonal

Of the endocrine system, sub-esophageal ganglion (SG) has been understood to play very vital role in the induction of diapause. It is generally accepted that egg diapause in silkworm is induced by an active principle secreted from a pair of large neurosecretory cells located in SG and are under the control of four other neurosecretory cells. This active principle is known as 'Diapause Hormone' (DH). Two forms of DH (DH-A and DH-B) have been identified which are neuropeptides having molecular weight of 2,000–3,000 and consist of 12–14 amino acids. The DH-A molecules contain 14 amino acids and two amino sugars, while DH-B contains the same amino acids but no amino sugars (Table 5.5). This reveals that amino sugar component of DH-A is apparently not essential for hormonal activity.

There are several differences between diapause and non-diapause eggs after oviposition. Diapause eggs showed high accumulation of 3-hydroxykynurenin, glycogen and ecdysteroids with reduced accumulation of cyclic guanosine monophosphate (GMP). Among these, 3-hydroxykynurenine and glycogen exhibit dramatic changes on the commencement of diapause, where 3-hydroxykynurenine is oxidized to ommochrome, resulting in the dark colouration of the diapause eggs and glycogen is mainly converted into sorbitol, which acts as an anti-freeze for the diapause embryo. DH also exerts an inhibitory effect on 'esterase 'A', an enzyme essential for mobilizing the yolk needed for completion of embryogenesis. Molecular analysis of DH action has demonstrated that DH directly induces trehalase gene expression in developing embryos, which eventually brings about hyperglycogenism in mature eggs, leading to sorbitol production at the onset of diapause. DH function is conceived to be the initial and essential reaction leading to the diapause-associated metabolism in silkworm eggs.

Table 5.5: Amino Acids Composition of Diapause Hormone–A (DH–A) and Diapause Hormone–B (DH–B) (After Yamashita and Hasegawa, 1985)

Amino Acid	*Diapause Hormone–A*	*Diapause Hormone–B*
Lysine	0.3	1.2
Arginine	0.8	1.0
Aspartic acid	1.3	1.3
Threonine	1.0	0.9
Serine	1.0	1.5
Gtutamic acid	2.1	2.2
Proline	3.0	2.4
Glycine	2.1	2.1
Alanine	2.2	2.1
Valine	1.7	1.0
Isoleucine	1.9	1.2
Leucine	3.0	3.0
Tyrosine	0.9	0.7
Phenylalanine	0.9	0.8
Tryptophane	–	1.0
Glucosamine	0.9	–
Galactosamine	0.9	–

Values represent mole ratios.

II. Biochemical Modulations

Since the first detailed study on the development of silkworm *(Bombyx mori)* eggs, embryological studies were confined especially in solving the practical problems such as identification of suitable stages for refrigeration of early embryos and the initiation, continuation and termination of diapause in order to develop an effective system for long-term cold storage of silkworm eggs. In these studies, morphological changes of embryos were used to assess their long-term survival. Moreover, the tolerance of eggs to cold storage varies with the stage of embryonic development besides genotypes because of adaptation phenomenon, metabolic changes and also genetic variations etc. and hence relatively recent studies on silkworm eggs becomes more concern with physiology, biochemistry and metabolic activity associated with termination of embryonic diapause for effective handling.

Biochemical Composition of Silkworm Egg

Newly laid eggs of silkworm are composed of protein (~ 10 per cent), lipids (~ 8.5 per cent), glycogen (~ 2.5 per cent), chorion (~ 18 per cent) and water (~ 60 per cent). More than 95 per cent of the total protein is yolk protein: vitellin (~ 40 per cent), 30 KD protein (~ 35 per cent) and egg-specific protein (ESP) (~ 20 per cent). These proteins are quite different from each other in their physiochemical and biological properties. Each protein exhibits a unique profile of degradation during embryogenesis, *viz.*,

- ☆ Egg-specific protein is utilized early and completely disappeared by the time of hatching.
- ☆ Vitellin begins to decrease at later stages of embryogenesis and some of it remains unutilized in hatched larvae. Thus, vitellin metabolism appears to be independent of the diapause phenomenon in silkworm (*Bobmyx mori)* eggs.
- ☆ 30 kD proteins are less utilized during embryogenesis.

ESP is degraded during embryogenesis by a process called hydrolysis catalyzed by trypsin like seryl protease and alters the carboxyl site of Lys^{114} and $Argenine^{21o}$ of ESP. The developmental increase in activity is due to increase at transcription of mRNA for this enzyme protein. Therefore, the utilization of ESP is a programmed event correlated with embryogenesis.

The second major component of egg is lipid (~ 8.5 per cent), which is composed of triglycerols (~ 80 per cent) and phospholipids (~ 20 per cent). Other constituents such as free fatty acids, mono and triglycerides are usually present in small quantities. Most of the metabolic energy (approximately 70 per cent of the total energy) is derived from the oxidation of triglycerol. The oxidation of lipids is advantageous for embryogenesis of terrestrial cleidoic eggs because large amount of metabolic water (1.07 g/g lipid) is released. Phospholipids are distributed as major component of yolk and used for formation of embryonic cells. Lipid is the major component consumed during diapause and the lipid concentration is higher in diapausing silkworm eggs than in non-diapausing eggs. Diapausing insects accumulate high level of glycerol and sorbitol led to the recognition that diapause is achieved by unique metabolic route that is not needed for non-diapausing eggs. Glycogen in silkworm eggs undergoes dramatic changes during diapause. In the diapause eggs, the following reversible reaction occurs with the initiation and termination of diapause:

$$\text{Glycogen} \leftrightarrow \text{Sorbitol} + \text{Glycerin}$$

With the initiation of diapause, the above reaction is right oriented while during termination, it is left oriented. The experiment with 14^C glycine showed that sorbitol is totally derived from glycogen, while glycerin is produced only when glycogen content reached the lowest level. About the mechanism of reversible reaction, following steps exists:

1. Glucose + NADPH ↔ Sorbitol + $NADP^+$
2. Glucose-6-p-fructose-6-p + NADPH ↔ Sorbitol-6-p + $NADP^+$
3. Glyceraldehydes + NADPH ↔ Glycerin + $NADP^+$
4. 2–OH acetone-p + NADH ↔ 2-p glycerin + NAD^+

Nucleic Acids Metabolism

In the silkworm eggs, diapause is decided during the maturation process of the egg in the ovary of pupal body. Therefore, there is a close relationship between diapause occurrence and metabolism of egg cells. In insects, nucleic acid is not only related to the expression of genes but also influence protein synthesis, cell division, growth and development. In univoltine genotypes, sub-esophageal ganglion if removed at early pupal stage, the female will lay non-diapausing eggs, while normal female laid diapause eggs. If the mature eggs inside the ovariole of above two groups taken out, it is found that DNA content of diapause eggs is 25.29 per cent lower than that of non-diapause eggs and RNA content of diapause mature eggs is 25.48 per cent less but the DNA/RNA ratio of these two groups were the same (Table 5.6). Hence, it is inferred that DNA content of mitochondria of diapause eggs is probably lower than non-diapause eggs.

Table 5.6: Relationship between Diapause and Metabolism in Mature Eggs of the Silkworm, *Bombyx mori* (*c.f.* Kurata *et al.*, 1980)

Characters	*Nature of Eggs*	
	Non-diapause	*Diapause*
DNA (μg/mg)	0.87 (100)	0.65 (74.71)
RNA (μg/mg)	18.84 (100)	14.04 (74.52)
Glycogen (per cent)	1.60 (100)	1.87 (116.88)
Lipids (per cent)	26.5 (100)	27.5 (103.77)
Oxygen Consumption ($mm^3/g/h$)	26.82 (100)	13.64 (50.19)

RNA plays a major role in protein metabolism and embryo morphogenesis. It is reported that mRNA carrying out the early morphogenesis information is synthesized during the egg formation process and is deposited in the cytoplasm of egg before fertilization. The entrance of sperm activates the egg and the mRNA in latent condition is activated first. During the pre-diapause stage, the DNA content in both non-diapause and diapause eggs are used very rapidly. But 48 hrs after oviposition, the DNA metabolism of these two types of eggs are quite different (Table 5.7). The DNA content of diapause eggs keeps constant after 48 hrs of oviposition is perhaps the biochemical pre-condition for stoppage of embryo morphogenesis and initiation of embryonic diapause. On contrary, the DNA content of non-diapause eggs continues to increase rapidly. In non-diapause eggs within 24 hours of oviposition, with the increase of DNA content, RNA content decreases. After 72 hrs of oviposition large amount of RNA is synthesized and accumulated in the non-diapausing eggs. This is closely related to the synthesis of protein during embryonic development. On contrary, in diapause eggs within three days of oviposition, the RNA content keeps constant though large quantity of DNA is synthesized and accumulated. Within 24 hrs of oviposition the RNA/DNA ratio in both the types of eggs (diapause and non-diapause) increases rapidly but after 48 hrs they reach the same level. Later the DNA/RNA ratio of non-diapause eggs increased rapidly.

Table 5.7: Nucleic Acids Metabolism During Early Embryonic Development of Diapause and Non-diapause Eggs in *Bombyx mori*

Embryo Development	**24*	**48*	***12*	***120*
DNA (μg/mg)				
Non-diapause	1.37	1.67	1.86	12.25
Diapause	1.00	1.49	1.54	1.58
RNA (μg/mg)				
Non-diapause	15.84	15.98	15.54	17.76
Diapause	14.27	14.33	14.00	13.58
DNA/RNA				
Non-diapause	0.086	0.104	0.120	0.127
Diapause	0.071	0.103	0.110	0.117

*Pre-diapaused period; ** Diapaused period.

Carbohydrate Metabolism

Carbohydrate metabolism is the main pathway of biochemical regulation of diapause in the silkworm. It is proved that induction; initiation, maintenance and termination of diapause are related to carbohydrate metabolism. Amount of glycogen accumulated in diapause eggs is 1.7 times higher than in non-diapause eggs. More than 90 per cent of carbohydrate accumulated in the silkworm diapause eggs is glycogen. The concentration of glycerin varies with genotype besides seasons from 32–40 mg/gm egg. Glycogen accumulated in the silkworm egg is from the glycogen stored in the fat body during pupal stage, which is converted into trehalose and is released into haemolymph and then absorbed by developing oocyte. The trehalase localizes in plasma membrane of vitellogenic follicles where haemolymph trehalose is hydrolyses into glucose to be taken up by oocytes. The glucose is immediately used to synthesize glycogen as a storage reserve, by which hyperglycogenia is induced in diapause eggs. Consequently, DH provides the metabolic state preparatory to diapause initiation. It has been demonstrated that enzyme involved in trehalose synthesis include trehalase, hexokinase, phosphoglucomutase, UDP-glucose–pyrophosphatase and glycogen synthetase. Of this trehalase is a membrane bound restriction enzyme existing on the surface of oocytes and follicle cells. The amount of trehalose in the pupal body fluid entering the ovary and the amount of glycogen synthesized in the egg is related to the trehalase activity. Thus, the increase of glycogen content in diapauses eggs depends on the increase of trehalase activity. At the initiation of diapause, the stored glycogen is converted into sorbitol by the activation of phosphorylase 'b' to 'a' under anaerobic condition and thus during the whole diapause stage, the glycogen content in the silkworm eggs keeps at a very low level but the sorbitol content keeps at a very high level. The pathway of glycogen to sorbitol is coupled to pentose phosphate pathway, through the supply of reducing power (NADPH). In these eggs, activity of glucose-6-phoaphate (G -6–P) dehydrogenase is about twice higher than that of phosphofructokinase and G–6–P amounts equivalent to about 10 per cent of the accumulated sorbitol are oxidized through pentose-phosphate pathway. The other fraction of G–6–P is eventually metabolized to glycerol. After the termination of diapause, sorbitol is again converted into glucose and the key enzyme involved in the activity is NAD-sorbitol dehydrogenase, which plays significant regulatory role in the metabolic cycle.

Accumulation of huge quantity of sorbitol in the diapausing eggs is generally regarded as a protecting agent of eggs from cold injury in winter. One of the remarkable difference between diapause and non-diapause eggs within few days of oviposition is that glycogen rapidly disappears in the diapausing eggs but remains at its initial level in the non-diapause eggs. The utilization of sorbitol is controlled by 'Nicotinamide Adenine Dinucleotide -Sorbitol Dehydrogenase' (NAD-SDH). Yaginuma *et al.* (1990) reported that incubation of diapause eggs at 25°C does not induce the activity of NAD–SDH until 90 days. Moreover, the activity appears at very low level between 90 -160 days. Similarly, chilling at 5°C for less than 30 days does not induce activity but 40 days period of chilling effectively stimulates activity. In contrast, 0.5°C never induces activity even with an exposure exceeding 250 days. Utilization of sorbitol in diapause eggs thus appears to be controlled by NAD–SDH activity through mediation of temperature. In non-diapause eggs, the sorbitol content increased on 2nd day of oviposition and then decreased rapidly.

Sorbitol Metabolism

Sorbitol Production

There is a significant biochemical difference at the time of oviposition between eggs destined to become diapause and non-diapause. The eggs developed in the presence of diapause hormone accumulate large quantity of glycogen. When oogenesis proceeded in the absence of diapause hormone, the laid eggs becomes non-diapause with less accumulation of glycogen and during embryogenesis, conversion of glycogen to sorbitol does not occur. Sorbitol is an ideal metabolite to assess the diapausing state of silkworm eggs. In diapause eggs, the initiation of sorbitol synthesis is correlated with the time when embryogenesis progressively slowed down. Conversion of glycogen to sorbitol at this stage is monitored by the enzyme 'glycogen phosphorylase-α activity. Conversion of glycogen phosphorylase-'β' to 'α' is being reported absent in non-diapause eggs under normal conditions.

Glycogen phosphorylase-β kinase responsible for conversion of 'β' to 'α' form is identified in silkworm eggs and is reported to have some similarities in kinetic properties to that in other insect tissues. There is no difference in their activity between diapause and non-diapause eggs. When newly laid non-diapause eggs are exposed to low temperature, sorbitol is accumulated and the extent of accumulation is closely dependent on the temperatures of exposure. In this case, glycogen is also used as an initial substrate for the formation of sorbitol. Thus, in non-diapause eggs, sorbitol production seems to be a biochemical adaptation against low temperature stress.

In diapause eggs, sorbitol began to decrease after continuous chilling for at least two months or by treatment with HCl on one month chilled eggs. The more advanced fall in sorbitol took place in eggs chilled for longer period. In all cases, sorbitol is stoichimometrically reconverted to glycogen. Thus, the conversion of sorbitol to glycogen is specific sign signally the breakdown of diapause in silkworm eggs. In the metabolic pathway from sorbitol to glycogen, NAD-sorbitol dehydrogenase, which catalyses the reaction from sorbitol to fructose is noticed to be the key enzyme. This activity is not being detected in eggs unchilled or chilled for less than two months and abruptly appears in eggs chilled for more than two months. The duration of diapause may be determined by analyzing the amount of oxygen required to prevent the decrease of sorbitol content in the eggs, which indicates that conversion of sorbitol to glycogen upon termination of diapause is closely related to the respiratory system via–GP cycle. Diapause termination is also closely related to the oxygen consumption during diapause period and that the activity of NAD–SDH enzyme is associated with the oxygen required for the termination of diapause. It is established that supply of oxygen to diapause eggs affect the maintenance of diapause period and when eggs are kept under anaerobic condition, the conversion of sorbitol to glycogen is delayed leading to delayed diapause termination.

Sorbitol Utilization

Diapausing silkworm eggs chilled at 5°C for at least three months when transferred to 25°C, the eggs resumed embryogenesis and hatches within two weeks. Soaking the pre-chilled diapausing eggs into hot solution can also bring about hatching. In these eggs, sorbitol content began to decrease after continuous chilling for at least two months or by treatment with HCl on one month chilled eggs. Greater fall in sorbitol content took place in eggs chilled for longer periods. Conversion of sorbitol to glycogen does not take place if eggs are kept continuously at 25°C even for more than six months. After diapause, all sorbitot in the eggs is converted into glycogen. During embryogenesis, the main carbohydrate consumed is glycogen. The glycogen-sorbitol-glycogen metabolic process is roughly the same as the process of diapause onset, maintenance and termination. In other words, the change of carbohydrate metabolism of the silkworm eggs is a close relative to the phenomenon of diapause. The sequence of major physiological events occurring in eggs of silkworm until the establishment of diapause is as:

One Day After Oviposition

- ☆ Optimum age for acid treatment to block diapause initiation; beginning of ommochrome formation in serosal cells.

- ☆ Beginning of glycogen decrease and maximum glycogen consumption.

Two Days After Oviposition

- ☆ Steep decline in oxygen consumption.
- ☆ Abrupt increase of glycogen phosphorylase-α activity.
- ☆ Decline in affinity of lysosomes in embryonic cells to acridine orange with supra vital staining; the affinity is lost during diapause.
- ☆ Arrest of increase in DNA content.

Three Days After Oviposition

- ☆ Arrest of nucleic acid synthetic activity in nuclei of embryonic cells and gradual decrease in yolk cells *in vitro*.

Four Days After Oviposition

- ☆ Arrest of mitotic activity in embryo.
- ☆ Continued steep decline in glycogen content and gradual accumulation of sorbitol and glycerol respectively.

Amino Acid and its Role

Significant changes in some free amino acids occurred during the initiation and termination of diapause. In particular, a sudden large increase in alanine content (about 50μ mol/gm eggs) occurred at the initiation of diapause (Suzuki *et al.*, 1984; Osanai and Yonezawa, 1986). Then alanine declines gradually with the increase of glutamate and especially proline. Content of proline that is always low during the initiation and maintenance of diapause, increased suddenly during the termination period indicating the conversion of alanine to glutamate and proline in diapausing eggs. In diapausing insects, a high concentration of free amino acids as well as of polyols and sugars serves to decrease the super cooling point (Somme, 1982). The super cooling point of *Bombyx mori eggs* is lower during diapause and hibernation (Suzuki *et al.*, 1983), and the increase of total amino acids and the accumulation of alanine or proline might be responsible for this effect. Proline serves as an energy source for later stages of embryonic life (Osanai and Yonezawa, 1986).

Chapter 6

PRESERVATION OF SILKWORM EGGS

Preservation refers to the management and protection of eggs from oviposition to hatching of the hibernating eggs. The success of silkworm crop depends on many factors like quality of seed, mulberry leaf, environmental factors, mode of rearing practices etc. Preservation of seed without affecting its quality and viability is an important factor in sericulture. The conditions under which eggs are preserved directly affect the hatching percentage, rearing performance and quality of cocoon and filament. The diapause and non-diapause characteristics of silkworm eggs are decided on the basis of the genetic characters and environmental conditions. In case of bivoltine silkworm, it is possible to make the eggs either to enter diapause or non-diapause through manipulation of incubation temperature. The cocoon quality of non-diapause eggs is usually inferior to that of silkworms that produce diapause eggs. Therefore, the preference of sericulturist is for diapausing eggs. These diapausing eggs can be made to hatch either through artificial treatment (hydrochlorization) or through preservation at low temperature for specific duration. The changes of external conditions have direct influence on the development of diapausing eggs. If preservation is not carried out properly, more eggs will die, hatching will not be uniform, larvae of the next generation will be weak and sericultural production will be affected.

However, the preservation of eggs at low temperature is in practice from the last ten decades, but it is not known who has first initiated the cold storage of silkworm eggs. Yokoyama (1973) stated that first large scale practice of preservation of silkworm eggs was carried out by Kisaburo Maeda who had used the natural caves situated at the footsteps of mount Fuji. The temperature in these caves ranged 0–4°C throughout the year.

The cold storage of silkworm eggs was improved by the use of refrigerator in 1902. Chotaro Yokota (1917) for the first time invented complex storage method. Mizuno (1920) suggested that safe duration of cold storage was dependent upon the storage of embryo, the temperature of storage and characteristics of silkworm variety. Considering these factors into account, several methods of storing of silkworm eggs for a long period without affecting the hatchability was reported (Yokoyama, 1973).

Conditions for Eggs Preservation

Climatic conditions plays very important role during preservation of eggs. Once eggs are laid, they undergo maturation division, fertilization and nuclear division and in about week's time, embryo enters a state of stable diapause. Eggs when laid are light yellow in colour, which gradually changes into reddish brown and after one and ½ days to two days, the colour continues to darken. On 5th day, the egg display specific colour fixed for the breed. During this period of development of eggs is very fast and rate of respiration is high. Therefore, eggs should be kept carefully to avoid shock, crush or rub. Again, the eggs should be preserved at 25°C temperature and any change in this optimum temperature will influence development of silkworm eggs.

If newly laid eggs are preserved to a temperature of more than 30°C, it affects the physiological activity and may result more unfertilized eggs and non-diapause eggs. The development below 20°C makes the diapause completion uncertain. Humidity should be maintained in between 75–85 per cent, as too much drying will degenerate the development of eggs. At the same time, excessive humidity is also harmful, as it will give rise to the development of fungus either directly on the eggs or on the egg card. This will increase the number of dead and unfertilized eggs.

Substances Harmful During Preservation of Silkworm Eggs

There are many substances, which are harmful to the development of silkworm eggs. The most likely to be encountered in the preservation of silkworm eggs are moulds, tobacco, insecticides, oils, alcohols, perfumes, paste, mercury, alkaline substances, bad smells etc.

Washing and Disinfection of Eggs

In order to get rid of the scaly hair, urine and pathogenic microorganisms attach to the surface of eggs and also to allow silkworm eggs to adhere properly to the surface of egg card in case of sheet eggs, the silkworm eggs must be washed properly and disinfected. The disinfection is done before transferring the eggs into cold storage. For disinfection, silkworm eggs are dipped in 2 per cent formaldehyde solution at 20°C temperature for 20 minutes by turning egg cards up and down during immersion to disinfect eggs evenly. After disinfection, the egg cards are washed in clean running water and hang up individually in shade to dry.

Preservation of Hibernating Eggs

Generally, bivoltine eggs are subjected to cold storage so that they can be used for rearing as and when required. The eggs produced in spring can be used for next spring or next autumn or next summer by adjusting the aestivation period after oviposition and cold storage. There is correlation between duration of aestivation (number of days preserved at normal room temperature) and cold storage as indicated in Table 6.1.

Two Step Cold Storage (Double Refrigeration or Intermediate Care)

Sometimes, silkworm eggs should be preserved in cold storage for rather a long period. In this case, procedure of two-step cold storage is to be adopted for better results. This means that silkworm eggs undergo two times preservation in cold storage. It is very well established that the embryo of silkworm eggs has got the corresponding tolerance limit for optimum cold storing temperature at each stage of embryonic development. Therefore, eggs are cold stored for certain period at specific stage of embryonic development

Table 6.1: Relationship Between Aestivation (25°C) and Cold Storage (5°C)

Aestivation Period (Days)	*1*	*3*	*5*	*10*	*20*	*40*	*60*	*80*	*100*
Shortest cold storage days	79	87	87	89	98	105	119	124	134
Longest cold storage days	139	165	168	178	193	202	196	161	145
Effective cold storage period (days)	61	79	82	90	96	98	78	38	12

Source: Narasimhanna, 1985.

to retain the vitality of embryo. The method of two-step cold storage of silkworm eggs is detailed below:

Step–1

Step 1 consists of eggs being cold stored at a particular embryonic developmental stage. The cold storage temperature should be around 5°C. Time limit for cold storage is around 90–100 days.

Step–2

Step 2 consists of releasing the eggs from the cold storage when they have reached the time limit of the first step and further preserved for 3 days at 15°C or 10 days at 10°C and a relative humidity of 85 per cent for allowing the embryos to reach the longest stage (Stage 15). The stage of development of silkworm eggs should be confirmed by embryo tests. When the development of silkworm eggs reaches to longest embryonic stage, they are subjected to second step of cold storage at 2.5°C for 10–40 days depending upon the need of hatching for undertaking brushing. Although, the limit for the second stage cold storage may be as long as 100 days but it is safer not to exceed for more than 80 days.

To identify stage of eggs suitable for refrigeration, embryo isolation is carried out and their characteristics are noted as indicated below and from which it can be inferred that while Stage 15 is suitable for double refrigeration, at Stage 16, eggs are to be brought at room temperature for incubation.

Stage–11

The formation of mesoderm starts at central portion of the embryo and extends to caudal and head region. The mesoderm shows regional differences in thickness and the arrangement appears to look like body segments.

Stage–12

After one day of release from cold storage at 15°C, the embryo appears slightly longer than Stage 11. The head and caudal lobes are longer and appearance of segments is slightly clear.

Stage–13

After 2 days of release at 15°C, the embryo appears longer than Stage 12. Mesodermal segments are clearly seen and depression in head lobe starts to appear.

Stage–14

After 2.5 days of release at 15°C, the embryo becomes slender and 18 segments can be clearly seen.

Stage–15

After 2.5–3 days of release at 15°C, the embryo is in its longest stage and occupies ¾ of the circumference of the egg. Depression in the head is clearly observed. It is actually the stage most suitable for double refrigeration.

Stage–16

After 3.5–4 days of release at 15°C, the embryo can be characterized by appearance of neural groove and appearance of body segments. This stage is not suitable for double refrigeration. At this stage, the eggs should be brought at room temperature for incubation.

The effective preservation in terms of number of days at various stages is presented in Table 6.2.

Table 6.2: Effective Cold Storage Duration of Diapause Eggs

Embryonic Stage	*Temperature of Cold Storage (°C)*			
	–2.5	*0*	*+2.5*	*+5*
Stage–11	170 days	150 days	150 days	120 days
Stage–12	120	120	70	70
Stage–13	70	70	70	70
Stage–14	70	100	90	40
Stage–15	60	60	100	30
Stage–16	60	60	80	30
Stage–17	30	30	30	30

Source: Mizuno.

Hibernation Schedule

In tropical countries, natural temperature available is not conducive for preservation of bivoltine silkworm eggs. Further in tropical conditions, climate is favourable for commercial silkworm rearing throughout the year. However, it is not possible to rear bivoltine silkworms through out the year to raise the seed cocoons. Therefore, bivoltine eggs are produced in favourable season and are released periodically as and when required. For this purpose, eggs are cold stored in cold storages, which are designed to have facilities for maintaining different required temperatures throughout the year with humidity of 70–80 per cent . Different hibernation schedules are recommended for various periods and are in-vogue to meet the seed supply programme throughout the year. During aestivation period, eggs are kept at 25 ± 1°C and 85 ± 5 per cent RH for various hibernation schedules. Eggs disinfected and dried properly are preserved under 4, 6 or 10 months schedule of hibernation depending on the demand of seed by rearers.

Schedule of Cold Storage for 4 Months

Bivoltine female moths allowed to lay eggs in oviposition room at 25 ± 1°C temperature and 85 ± 5 per cent humidity are examined next day for disease freeness. Healthy eggs are washed in 2 per cent formalin solution for 10–15 minutes, and then washed in running water to remove excess of formalin. These eggs are then hanged, shade dried and transferred to cold storages, where they are preserves at 25°C for 10 days. These eggs are transferred at 20°C for 2 days; 15°C for 2 days; 10°C for 3 days, 5°C for 50 days and 2.5°C for 50 days. They are then released from cold storages and are subjected to 15°C for 6 hrs and incubated at 25°C temperature and 70–80 per cent humidity (Table 6.3).

Table 6.3: Four (4) Months Hibernation Schedule

Preservation temperature (°C)	25	20	15	10	5	2.5	15	25
Preservation duration (days)	10	2	2	3	50	50	6 hr	Incubation till hatching

Schedule of Cold Storage for 6 Months

Female moths are allowed to lay eggs in oviposition room where a temperature of 25 ± 1°C and humidity of 85 ± 5 per cent is maintained. Most of the eggs are laid in dusk. Next day moths are examined for disease freeness and healthy eggs are washed in 2 per cent formalin solution for 10–15 minutes. Then these eggs are washed in running water to remove excess formalin. The eggs are dried in shade and kept at 25 ± 1°C temperature and humidity of 85 ± 5 per cent for 20 days. The eggs are then transferred to cold storage. At cold storage, the eggs are kept at 20°C for 15 days; 15°C for 10 days; 10°C for 10 days; 5°C for 50 days and 2.5°C for 60 days. The eggs are then transferred to 15°C for 1 day and later released for incubation at 25°C and 70–80 per cent humidity (Table 6.4).

Table 6.4: Six (6) Months Hibernation Schedule

Preservation temperature (°C)	25	20	15	10	5	2.5	15	25
Preservation duration (days)	20	15	10	10	50	60	1 day	Incubation till hatching

Schedule of Cold Storage for 10 Months

Two-step refrigeration is followed for preservation of eggs for long duration of 10 months. Female moths are allowed to lay eggs in oviposition room where a temperature of 25 ± 1°C and humidity of 85 ± 5 per cent is maintained. They are washed in 2 per cent formalin for 10–15 minutes and then transferred in running water to remove excess of formalin. These eggs are dried in shade and then to cold storage where they are preserved at 25°C for 50 days: 20°C for 40 days; 15°C for 25 days; 10°C for 25 days; 5°C for 60 days and 2.5°C for 55 days. These eggs preserved at 2.5°C for 55 days are released for intermediate care at 15°C for 3–4 days and then cold stored again at 2.5°C for 30–40 days. They are released for incubation at 25°C and 70–80 per cent humidity after completion of specific period of cold storàge (Table 6.5).

Reddy *et al.* (2005) reported different hibernation schedule based on aestivation period (3 days aestivation to 60 days aestivation period) and stated that silkworm eggs can be preserved in cold storages for longer period even up to 335 days and release them as per requirement. Hibernation schedule from 3 days to 60 days aestivation period and their cold storage to the maximum of 335 days as recommended by Reddy *et al.* (2005) is furnished in Table 6.6 to Table 6.10.

Reddy *et al.* (2005) have also measured the glycogen level in two pure bivoltine breeds and two hybrids at different days of oviposition and or aestivation (Table 6.11).

Table 6.5: Eight (8) to Ten (10) Months Hibernation Schedule

Preservation temperature (°C)	25	20	15	10	5	2.5	5	15	2.5	5	15	25
Preservation duration (days)	50	40	25	25	60	55	1	3–4	30–40	1	1	Incubation till hatching

Table 6.6: Hibernation Schedule Under 3 Days Aestivation

Temperature (°C)	*25*	*20*	*15*	*10*	*5*	*2.5*	*5*	*15*	*2.5*	*5*	*15*	*Total Duration (Days)*
Preservation	3	1	1	1	30	30	–	–	–	1	3	70
duration	3	1	1	1	40	40	–	–	–	1	3	90
(days/hrs)	3	1	1	1	50	50	–	–	–	1	3	**110**
	3	1	1	1	60	60	–	–	–	1	3	**130**
	3	1	1	1	70	70	–	–	–	1	3	**150**
	3	1	1	1	80	80	–	–	–	1	3	**170**
	3	1	1	1	90	90	–	–	–	1	3	**190**
	3	1	1	1	100	100	–	–	–	1	3	**210**
	3	1	1	1	100	100	1	3	20	1	3 hr	**231**
	3	1	1	1	100	100	1	3	30	1	3 hr	241
	3	1	1	1	100	100	1	3	40	1	3 hr	251
	3	1	1	1	100	100	1	3	50	1	3 hr	261

Safe Period.

Table 6.7: Hibernation Schedule Under 10 Days Aestivation

Temperature (°C)	*25*	*20*	*15*	*10*	*5*	*2.5*	*5*	*15*	*2.5*	*5*	*15*	*Total Duration (Days)*
Preservation	10	4	4	4	60	30	–	–	–	1	3	96
duration	10	4	4	4	70	40	–	–	–	1	3	116
(days/hrs)	10	4	4	4	80	50	–	–	–	1	3	**136**
	10	4	4	4	90	60	–	–	–	1	3	**156**
	10	4	4	4	100	70	–	–	–	1	3	**176**
	10	4	4	4	100	80	–	–	–	1	3	**196**
	10	4	4	4	100	80	1	3	10	1	3 hr	**217**
	10	4	4	4	100	80	1	3	20	1	3 hr	**227**
	10	4	4	4	100	80	1	3	30	1	3 hr	**237**
	10	4	4	4	100	80	1	3	40	1	3 hr	**247**
	10	4	4	4	100	80	1	3	50	1	3 hr	257
	10	4	4	4	100	80	1	3	60	1	3 hr	267

Safe Period.

Table 6.8: Hibernation Schedule Under 20 Days Aestivation

Temperature (°C)	*25*	*20*	*15*	*10*	*5*	*2.5*	*5*	*15*	*2.5*	*5*	*15*	*Total Duration (Days)*
Preservation	20	4	4	4	60	30	–	–	–	1	3	106
duration	20	4	4	4	70	40	–	–	–	1	3	126
(days/hrs)	20	4	4	4	80	50	–	–	–	1	3	**146**
	20	4	4	4	90	60	–	–	–	1	3	**166**
	20	4	4	4	100	70	–	–	–	1	3	**186**
	20	4	4	4	100	80	–	–	–	1	3	**206**
	20	4	4	4	100	80	1	3	10	1	3 hr	**227**
	20	4	4	4	100	80	1	3	20	1	3 hr	**237**
	20	4	4	4	100	80	1	3	30	1	3 hr	**247**
	20	4	4	4	100	80	1	3	40	1	3 hr	**257**
	20	4	4	4	100	80	1	3	50	1	3 hr	267
	20	4	4	4	100	80	1	3	60	1	3 hr	277

Safe Period.

Table 6.9: Hibernation Schedule Under 40 Days Aestivation

Temperature (°C)	*25*	*20*	*15*	*10*	*5*	*2.5*	*5*	*15*	*2.5*	*5*	*15*	*Total Duration (Days)*
Preservation	40	20	15	10	50	–	–	–	–	1	3	139
duration	40	20	15	10	50	20	–	–	–	1	3	159
(days/hrs)	40	20	15	10	60	30	–	–	–	1	3	179
	40	20	15	10	70	40	–	–	–	1	3	**199**
	40	20	15	10	80	50	–	–	–	1	3	**219**
	40	20	15	10	90	60	–	–	–	1	3	**239**
	40	20	15	10	100	70	–	–	–	1	3	**259**
	40	20	15	10	100	80	1	3	10	1	3 hr	**280**
	40	20	15	10	100	80	1	3	20	1	3 hr	**290**
	40	20	15	10	100	80	1	3	30	1	3 hr	**300**
	40	20	15	10	100	80	1	3	40	1	3 hr	310

Safe Period.

Table 6.10: Hibernation Schedule Under 60 Days Aestivation

Temperature (°C)	*25*	*20*	*15*	*10*	*5*	*2.5*	*5*	*15*	*2.5*	*5*	*15*	*Total Duration (Days)*
Preservation	60	30	10	10	50	–	–	–	–	1	3	164
duration	60	30	10	10	50	20	–	–	–	1	3	184
(days/hrs)	60	30	10	10	60	30	–	–	–	1	3	204
	60	30	10	10	70	40	–	–	–	1	3	**224**
	60	30	10	10	80	50	–	–	–	1	3	**244**
	60	30	10	10	90	60	–	–	–	1	3	**264**
	60	30	10	10	100	70	–	–	–	1	3	**284**
	60	30	10	10	100	80	1	3	10	1	3 hr	**305**
	60	30	10	10	100	80	1	3	20	1	3 hr	**315**
	60	30	10	10	100	80	1	3	30	1	3 hr	**325**
	60	30	10	10	100	80	1	3	40	1	3 hr	335

Safe Period.

Table 6.11: Glycogen Level During Different Days of Aestivation

Days After Oviposition	*Glycogen Content (mg/g of Eggs)*			
	NB_4D_2	*CC_1*	*NB_4D_2 x PM*	*CC_1 x NB_4D_2*
1	37.03	38.21	34.96	38.73
2	25.35	27.33	24.47	28.14
3	23.33	23.81	20.05	22.87
5	21.56	22.22	16.30	18.85
10	14.83	12.96	12.30	13.99
20	14.39	13.37	11.46	12.82
25	11.19	12.12	08.15	11.48
30	10.78	08.24	08.62	09.39
35	08.18	08.38	07.35	05.60
40	07.37	07.82	06.55	04.90
45	06.38	06.02	06.69	04.74
50	06.43	06.54	06.60	05.43
55	06.00	06.20	06.43	05.25
60	06.22	06.26	06.29	05.20

Normally silkworm eggs are not stored for more than one year and earlier temperature regimes did not use zero or below zero Celsius temperature. Relatively recently, Takeshi Shirota of the Silk Science Research Institute, Amimachi, Ibaraki of Japan has reported a method of two year preservation of silkworm eggs. He studied the long-term preservation of silkworm eggs using silkworm hybrids and found the optimum temperature for egg preservation to be from -5 to 0°C. The optimum temperature regime involved the following procedure:

Preservation for 30 days at 25°C → 202 days at -5°C → 200 days at 0°C → 61 days at 2.5°C → 5 days at 5°C → 5 days at 10°C → 61 days at 0°C → 1 day at 2.5°C → 1 day at 5°C → 4 days at 10°C → 3 days at 15°C (intermediate care) → 62 days at 2.5°C → 1 day each at 5, 10, 15°C and then incubation for 12–13 days at 25°C. The total duration thus calculated will be 651 days including incubation. However, mean hatchability observed by him was only 40 per cent. This method may be useful in germplasm maintenance but not in commercial rearing methods, when hatching above 90 per cent is required at any given period of time.

Cold Storage of Acid Treated Eggs

Silkworm eggs, which are acid treated by the general method at 46°C of pretreated HCl of 1.075 specific gravity, can be cold stored for 10–20 days only. It is always better and suggested to avoid cold storage of eggs after acid treatment. However, if it is inevitable, the eggs preserved for 12 hrs at 25°C following acid treatment are transferred to 15°C for 6 hrs. They are then transferred to 5°C up to 20 days. Cold storage duration should be kept short as far as possible but in no circumstances it should be beyond 20 days. The optimum time for cold storage of acid treated eggs is when the embryos have developed appendages. If cold storage is delayed, it leads to death of embryos resulting in poor hatching. If cold storage of acid treated eggs is carried out after 48 hrs of acid treatment, the preservation duration must be kept less than 10 days.

Acid Treatment of Eggs After Cold Storage

In order to postpone hatching over a period of time but not as long as regular hibernation or not as early as by acid treatment, a technique of chilling the eggs and acid treatment in combination is followed (Table 6.12).

Table 6.12: Schedule of Hatching at Different Periods

*Duration of Preservation***	*Type of Preservation*	*Procedure*
10–30 days	Cold storage of acid treated eggs	Acid treatment after 20 hrs of oviposition →15°C for 6 hrs → 5°C up to 20 days.
40–50 days	Short term chilling	30–35 hrs at 25°C after oviposition → 15°C for 6 hrs → chilled at 5°C for 30–40 days → release and acid treated.
50–60 days	Long term chilling	
	i) 40-50 days chilling	40–50 hrs at 20°C after oviposition → 15°C for 6 hrs → chilled at 5°C for 40–50 days → released and acid treated.
	ii) 60-70 days chilling	40–50 hrs at 25°C after oviposition → 15°C for 6 hrs → chilled at 5°C for 40 days → transferred to 2.5°C for 20–30 days → released and acid treated.

** Incubation period included

If the eggs are acid treated within an hour of release from the cold storage, the sudden change in temperature adversely affect the eggs leading to poor hatchability. At the same time, if the eggs are acid treated after 3 hours of release from cold storage, the desired result is not obtained as hatching is adversely affected. Therefore, the eggs after being released from cold storage must be subjected to acid treatment within 1–3 hrs. For the treatment, hydrochloric acid having specific gravity of 1.100 at 15°C pre-heated to a temperature of 48°C is used. Duration of acid treatment is usually 5–6 minutes but is variable according to the silkworm breeds/hybrids.

(*a*) Short-Term Chilling and Acid Treatment

Eggs laid at 25 ± 1°C temperature and humidity of 85 ± 5 per cent is preserved for 30–35 hrs after ovipositions are transferred at 15°C for 6 hrs. By this time, the eggs turn into reddish brown colour. These eggs are then cold stored at 5°C temperature and 70–80 per cent humidity for 30–40 days. They are released from cold storage after completion of scheduled duration and kept at room temperature (25°C) for 1–3 hrs before performing acid treatment at 48°C for 5–6 minutes by dipping in preheated HCl solution of 1.100 specific gravity. This is known as short term chilling.

(*b*) Long-Term Chilling and Acid Treatment

40–50 Days Chilling

The eggs laid at 25 ± 1°C temperature and humidity of 85 ± 5 per cent is preserved in the same condition for 40–50 hrs. By this time, the colour of eggs changes to reddish brown and embryos have reached the spoon shape stage. These eggs are stored at 15°C for 6 hrs and then transferred to cold storage at 5°C for 40–50 days. These eggs after completion of specific duration are released from cold storage and kept at 25°C for1–3 hrs. These eggs are subjected to hot acid treatment at 48°C for 5 minutes in HCl solution of 1.100 specific gravity.

60–70 Days Chilling

In case the hatching of the eggs are to be delayed beyond 60 days, the eggs within 40–50 hrs after oviposition preserved at 25 ± 1°C temperature and humidity of 85 ± 5 per cent are transferred to cold storage at 5°C for 40 days. Later these eggs are transferred and subjected to 2.5°C temperature for 20–30 days and then brought back to room temperature (25°C) and treated with HCl solution of 1.100 specific gravity at 15°C pre-heated to 48°C for 5–6 minutes. This is known as long term chilling.

Calculation of Age of Embryo

Zero Hour Embryo

When the pairing time is 7 AM–10 AM, the zero hour will be 6 PM–8 PM on the same day.

24 Hours Embryonic Age

When the pairing time is 7 AM–10 AM, 8 AM on the next day is considered as 24 hours of embryonic age.

36 Hours of Embryonic Age

When the pairing time is 7 AM–10 AM, 8 AM on the 3rd day of oviposition is considered as 36 hours embryonic age.

These embryonic ages are considered at optimum conditions of temperature and humidity in the oviposition room. However, if the optimum required conditions are not maintained, the embryonic age are considered depending on the prevailing conditions.

Precautions During Preservation of Eggs

- ☆ Aestivation of eggs at 25 ± 1°C temperature and 85 ± 5 per cent Relative Humidity (RH) for required duration is necessary to establish diapause.
- ☆ Diapause eggs should pass through moderately low temperature (20°C, 15°C and 10°C) before the start of cold storage at 5°C or 2.5°C to avoid cold injury/shock to the eggs.
- ☆ While releasing eggs from the cold storage for incubation, they must pass through intermediate temperature of 15°C to reduce high temperature injury.
- ☆ To terminate diapause effectively, it is necessary to cold store the eggs at 5°C for 50–60 days.
- ☆ For longer duration of cold storage (double refrigeration) at 2.5°C, it is necessary to provide intermediate care at 15°C for 3 days in order to prolong hibernation schedules.
- ☆ Exposure of hibernating eggs to 15°C for 3 days after release from cold storage, not only avoid a sharp change

in temperature but also prepare the younger embryos to develop uniformly in order to obtain single day uniform hatching in hibernating eggs of bivoltine.

☆ Since, termination of diapause needs sufficient oxygen from the time, it is necessary to provide sufficient space in the cold storage for developing embryos.

☆ Compactness during preservation in cold storages leads to irregular hatching and poor quality of crop and hence should be avoided.

Damage of Silkworm Eggs During Preservation

Dead Blue-Eggs

The eggs during incubation develops up to blue stage and then die due to cumulative effect of various conditions prevailing during mounting, preservation of pupae, artificial hatching and incubation. In this case, larvae develop completely inside the egg but it does not have the capacity to break the eggshell and as a result dies inside the eggshell. Changes in the temperature during cold storage, duration and temperature prevailing during intermediate care, duration and temperature of water during washing, degree of stimulation through water or acid treatment, effect of chemicals besides unfavourable environmental conditions during incubation and handling are responsible for eggs to die at blue stage.

Dead Coloured-Eggs

Usually refers to the eggs, which are unfertilized and red in colour. Due to improper method of preservation and cold storage, the viability of eggs is lost even before the pigmentation stage and the eggs die. This type of eggs are also caused due to improper handling of eggs, improper washing of eggs, rapid activation, insufficient diapause, poor variety of breeds, improper acid treatment besides due to affect of chemicals.

White Fluffy-Eggs

These types of eggs are usually caused when they are transferred from high temperature to low temperature without passing through intermediate temperature. These types of eggs are also caused due to too much drying after washing or acid treatment soon after transportation. If the effect is very mild, the development proceed up to revolving stage of eggs and sometimes larvae hatches out from the eggshells and shows normal larval development.

Unfertilized Eggs

Unfertilized eggs are caused due to unfavourable environmental conditions during mounting, improper handling of moths during mating and egg laying. Unfertilized eggs are observed in large numbers among the eggs laid at the end of oviposition as compared to the eggs laid at the beginning of oviposition. Acid treatment for longer period also causes mortality of eggs, which appears to be unfertilized. Unfertilized eggs do not form any colour and if these eggs are pressed, they do not rupture but makes sound.

Abnormal Hatching

In a laying, though eggs appear like diapausing eggs but hatch within few days after oviposition *i.e.* during aestivation period of diapausing eggs. This is caused due to improper management during pupal preservation especially when the temperature is not maintained properly.

Abnormal Larvae

Abnormal larvae are caused due to improper acid treatment and cold storage of eggs. Though, it is considered to be a genetic disorder but also occurs due to abnormal temperatures and influence of HCl solution. Besides above abnormalities, irregular hatching and mortality of new born larvae, light yellow coloured eggs etc. are also occurring during preparation and preservation of eggs.

Transportation of Silkworm Eggs

Various commodities are transported daily by various means, but egg transportation has to be carried out with all care and diligence in all conditions, which foster continued healthy growth of silkworm. It is emphasized that maintenance of recommended temperature and humidity is very crucial during transportation of eggs from cold storages/grainages to incubation centers. The transportation of silkworm eggs has to be carried out by providing optimum environmental conditions of temperature (25 ± 1°C) and humidity (75 ± 5 per cent) to exploit the full potential of the breed/hybrid. If eggs are exposed to unfavourable conditions for longer duration, it would have a tremendous impact on the developing embryos which may cause irreparable damage. Once, the damage is inflicted, the silkworm eggs being a closed system can not circumvent the imposition of nature and the penalty is imminent. The implications would be mortality of the developing embryos, poor and irregular hatching, weaker hatched larvae and loss of weight in the newly hatched larvae and consequent death in 2–3 days after brushing, susceptibility to diseases and increased early stage loss. Temperature of 25 ± 1°C even though ideal during transportation of eggs but it becomes dangerous when RH is 35 per cent or less. It not only affects hatching but also retards growth and extend incubation period. Under acute hostile conditions, the embryo could collapse in the mid-course of development or may extend its lease of life up to pin-head/blue egg stage and die. On the other hand, if RH is too high *i.e.* 90 per cent and above, it leads to retention of water, a product of oxidation of nutrients and poisons the embryos. Such embryos develop into weaker larvae and become easily susceptible to diseases. Therefore, silkworm eggs has to be carried out by providing the nearest optimum temperature and RH. It is recommended to transport eggs during cooler hours of the day using egg carrying bag or box in order to provide required temperature and RH equilibrium.

Chapter 7

ARTIFICIAL HATCHING OF EGGS

Based on the number of generations in a year, silkworm eggs are classified as univoltine–the eggs of which hatches once a year, bivoltine–the eggs of which hatches twice a year and polyvoltine–the eggs of which hatches several times a year and thus resulting into one, two and many generations in a year respectively. These eggs are classified as diapausing and non-diapausing eggs. Bivoltine and univoltine enters into diapause within 40–50 hrs of egg laying at 25°C. The eggs when laid are yellow in colour. Gradually they change into light brown and then to purple brown. Under normal conditions, these diapausing silkworm eggs remain in diapause for a year after being laid and will not hatch until either given artificial stimulation or subjected to cold storage for a sufficient period. The mother moth in univoltine/bivoltine forms releases a hibernating substance. This diapausing/hibernating substance when released, arrest development of embryos and thus embryos will not develop even under optimum environmental conditions. However, after eggs are laid, if they are subjected to an artificial treatment at an appropriate age, stage and time, stimulate further growth without allowing them to enter into diapause. This *'artificial method of stimulating silkworm eggs for hatching is known as artificial hatching'*.

Characteristics of multivoltine and bivoltine eggs are summarized in Table 7.1.

Methods of Artificial Hatching of Silkworm Eggs

There are several methods of artificial hatching which can be broadly classified into physical and chemical methods -

(A) Physical Methods

- ☆ Low temperature stimulation (chilling)
- ☆ Hot water treatment
- ☆ Stimulation by friction
- ☆ Electrical stimulation
- ☆ Stimulation by exposing to sunshine
- ☆ Stimulation by high atmospheric pressure
- ☆ Artificial over wintering
- ☆ Stimulation by ultraviolet/supersonic treatment

(B) Chemical Methods

- ☆ Hydrochloric acid treatment
- ☆ Nitric acid treatment
- ☆ Sulphuric acid treatment
- ☆ Enzyme treatment
- ☆ Ozone treatment
- ☆ Sodium chloride solution treatment
- ☆ Per chloride treatment

Tanaka (1964) stated that Verson was apparently the first discovered artificial hatching by electrical stimulation. Later Araki and Miura succeeded in using a high voltage electrical source and stated that negative electrical discharge is more effective than positive electrical discharge. With the discovery of this method large electrical hatching machine using negative electrical discharge by intermittent direct current of 25000–50000 volts were fabricated. Terni and others discovered friction method. In this method eggs laid on cards when brushed strongly for 5–6 minutes by a hard brush, 40–50 per cent eggs laid will hatch as quoted by Tanaka (1964).

Table 7.1: Characteristics of Multivoltine and Bivoltine Silkworm Eggs

Multivoltine	*Bivoltine*
(*a*) Eggs are non-pigmented and non-hibernating.	(*a*) Eggs are pigmented and hibernating.
(*b*) Eggs hatches within 10–12 days of oviposition.	(*b*) Eggs do not hatch naturally after 10 of oviposition.
(*c*) Acid treatment is not required to stimulate hatching.	(*c*) Acid treatment is required to stimulate hatching.
(*d*) Flat card/sheet egg method is applied for egg production.	(*d*) Eggs are mostly produced by loose production method.
(*e*) Sexes are separated at moth stage for production of hybrid.	(*e*) Sexes are separated at pupal stage for production of hybrid.
(*f*) Eggs cannot be preserved for more than 20 days at 5°C.	(*f*) Eggs can be preserved up to 6-10 months under different hibernation schedules.

Kui and Van reported that when hibernating eggs are dipped into hot water at 51–56°C for 3–15 seconds about 10 hrs after egg laying, results in 90 per cent hatching. Gowda *et al.* (1998) stated that eggs treated with hot water at 55°C for 8 seconds duration resulted in good hatchability. They observed large number of dead eggs with the increase in temperature or duration of treatment. Ishii reported that when eggs were soaked in 3 per cent hydrogen peroxide at 40–47°C for 7–10 minutes on the second day of egg laying; 100 percentages of them hatched. Rollet reported exposing of eggs to high atmospheric pressure. Moreover, the artificial hatching method should have high and uniform hatching rate, simple equipments, easy to handle besides high work efficiency and economical in use. Taking into consideration of these facts, immersion of eggs in hydrochloric acid solution have been established most economical and safe, to obtain higher hatchability. This technique is now being widely practiced for the treatment of bivoltine eggs.

Appliances for Acid Treatment

- ☆ Acid treatment bath
- ☆ Specific gravity meter
- ☆ Measuring cylinder (500–1 000 ml capacity)
- ☆ Clock (A stopwatch with second's hand)
- ☆ Acid frame and acid container (for loose eggs)
- ☆ Water tank for removal of acid
- ☆ Drying apparatus.

If acid treatment bath is not available, then following apparatus will also be required:

- ☆ Hydrochloric acid container
- ☆ Heater
- ☆ Thermometer

Preparation of Hydrochloric Acid Solution

(*a*) Hydrochloric Acid

There are pure and industrial varieties of hydrochloric acid (HCl). The saturated solution of pure hydrochloric acid is colourless, transparent, dissolves easily in water, and vaporizes immediately into a thick white 'fog' when it comes in contact with air. It is strong stimulant, corrosive and at 15°C contains 42.09 per cent hydrogen chloride. The specific gravity of pure hydrochloric acid is 1.212. However, it is better to use pure form of HCl solution for acid treatment but it is quite expensive. Industrial HCl is light brown or yellow in colour. It contains 30 per cent hydrogen chloride and has specific gravity of 1.180. It is suggested to use always commercial grade chemically pure hydrochloric acid with 1.15 to 1.18 specific gravity and 30 per cent hydrogen chloride. Commercial grade acid used must be free from impurities like mercury, nitrates and fluorides etc. Otherwise there would be problem of burning/poisoning of embryos/non-uniform change of egg colour which may create problem at the time of hatching.

There is positive correlation between specific gravity and concentration of HCl. In other words, if specific gravity increases, the concentration of HCl solution increases and vice-versa. The specific gravity of HCl solution also varies with the temperature of the acid and there is negative correlation between the two. In other words, if the temperature of HCl solution is higher, the specific gravity will be lower (Table 7.2).

Table 7.2: Ready Reckoner of Specific Gravity for the Preparation of Hydrochloric Acid Solution

Temperature	*Specific Gravity (°C)*	*Temperature (°C)*	*Specific Gravity*
15	1 .0750	27	1.0710
16	1.0746	28	1.0704
17	1.0743	29	1 .0708
18	1.0739	30	1 .0697
19	1.0736	31	1 .0691
20	1 .0732	32	1.0690
21	1.0729	33	1.0687
22	1.0725	34	1 .0683
23	1 .0722	35	1.0680
24	1.0718	40	1 .0666
25	1.0715	45	1 .0649
26	1.0711	50	1 .0632

From 15°C, when temperature increases or decreases, the co-efficient of specific gravity of HCl acid solution varies as follows (Table 7.3).

Table 7.3: Relationship Between Specific Gravity and Co-efficient of Variation

Range of Specific Gravity	*Variation of Co-efficient of Specific Gravity*
1.065–1.075	0.0003
1.080–1.100	0.0004
1.105–1.120	0.0005

Approximate specific gravity of HCl solution can be estimated with the help of temperature and variable co-efficient. The concentration of HCl at saturation is 40 per cent and density (specific gravity) at 20°C is 1.198 (Table 7.4). Concentrated solution of HCl contains 28–36 per cent hydrogen chloride by weight, which can be modified by addition of water.

Table 7.4: Relationship Between Specific Gravity and Concentration of HCl

Sl.No.	*Specific Gravity of HCl at 20°C*	*Concentration of HCl (per cent)*
1	1.198	40
2	1.195	39
3	1.190	38
4	1.180	36
5	1.170	34
6	1.160	32
7	1.150	30
8	1.140	28
9	1.130	26
10	1.120	24
11	1.110	22
12	1.100	20
13	1.095	19

Contd...

Table 7.4–Contd...

Sl.No.	Specific Gravity of HCl at 20°C	Concentration of HCl (per cent)
14	1.090	18
15	1.085	17
16	1.080	16
17	1.075	15
18	1.070	14

(*b*) Concentration of Hydrochloric Acid

There are two methods of diluting the original solution of HCl to obtain the desired concentration of solution for acid treatment. They are:

Dilution Based on Concentration

While preparing the solution of desired concentration by adding water to the stock solution, the amount of water to be added can be calculated on the basis of the following formula:

$$A = \frac{B - C}{C}$$

where,

A: Quantity of water to be added

B: Concentration of stock solution

C: Concentration of desired solution

For example, by using the stock solution of 38 per cent concentration of industrial HCl, 2 per cent concentrated solution of desired strength can be prepared in the following way:

$$A = \frac{B - C}{C} = \frac{38 - 2}{2} = 18$$

Ratio, 18:1

Therefore, eighteen parts of water should be added to one part of stock solution to prepare 2 per cent hydrochloric acid solution from 38 per cent stock solution of industrial grade.

Dilution Based on Specific Gravity

When preparing solution of HCl acid of desired gravity on the basis of specific gravity, the quantity of stock solution to be used can be estimated by the following formula:

$$A = \frac{B-1}{C-1} \times 100$$

where,

A: Quantity of stock solution required
B: Desired specific gravity
C: Specific gravity of stock solution

For example, in order to prepare the acid solution of 1.075 specific gravity by using the stock solution of 1.18 specific gravity of HCl, the quantity of stock solution to be used will be:

$$A = \frac{B-1}{C-1} \times 100, \textit{ i.e. } A = \frac{1.075-1}{1.180-1} \times 100 = 41.66$$

Table 7.5: Ready Reckoner for Preparing One Liter Acid of Desired Specific Gravity (Quantity in ml)

Specific Gravity of Available HCl	1.075 HCl		1.100 HCl		1.110 HCl	
	Water	Acid	Water	Acid	Water	Acid
1.150	500	500	333	667	267	733
1.155	516	484	355	645	290	710
1.160	531	469	375	625	312	688
1.165	545	455	394	606	333	667
1.170	559	441	412	588	353	647
1.175	571	429	429	571	371	629
1.180	583	417	444	556	389	611

Therefore, to prepare 100 liters of HCl solution of 1.075 specific gravity, 41.66 liters of stock solution of specific gravity 1.180 and 58.34 liters of water (100 minus 41.66) will be required. When the acids get turbid or once in three days or after treating 5 kgs of eggs, filter the acid and reuse. Every time before the treatment, the specific gravity of the solution should be measured to ensure to make it to the desired level.

Artificial Hatching by Acid Treatment

Artificial hatching is usually carried out before the appearance of hibernating characters to block the hibernating tendency and stimulate the embryos for further development instead of allowing them to enter into diapause. There are two methods of acid treatment:

- ☆ Hot acid treatment (common acid treatment)
- ☆ Cold acid treatment (room temperature treatment).

However, in these two types of acid treatment, the concentration and temperature of HCl acid, the specific gravity of acid and duration of acid treatment may vary but in principle both the method is one and the same and is being carried out with the same purpose.

Hot Acid Treatment

Time of Acid Treatment

Acid treatment is to be conducted at the appropriate age and stage of development of silkworm eggs. The best time considered is between the fusion of amnionic fold and formation of embryo which is in between 16–30 hrs of egg laying at 25°C. For safer point of view, the eggs between the age groups of 20–24 hrs are ideal for acid treatment. Therefore, if time of egg laying is taken as 20 hrs (8 P.M. night) (by the time when most of the eggs are laid), then the acid treatment should be conducted between 16–20 hrs the following day. However, if temperature at the time of egg laying and preservation is high, acid treatment should be conducted slightly ahead and a little later if preservation temperature is low. Since, the rate of embryonic development differs according to the preservation temperature of eggs after egg laying, the time of acid treatment is decided on the basis of time lapse and egg preservation temperature. Temperature range between 27–30°C does not have any harmful affect on egg development; moreover, preservation temperature above 30°C after egg laying is harmful and should be avoided in order to achieve higher hatchability, healthy larvae and stable cocoon crop.

Concentration of HCl Acid Solution

The concentration of HCl solution for hot acid treatment should be 15 per cent with 1.075 specific gravity.

Temperature of HCl Solution

HCl acid solution heated up to 46°C (115°F) is used for hot acid treatment of silkworm eggs. The temperature of the solution should not vary ± 0.5°C.

Duration of Acid Treatment

The 5 minutes duration of treatment have been considered adequate with HCl acid solution heated up to 46°C having 1.075 specific gravity. Acid should be heated indirectly. Temperature of acid should be checked from time to time during treatment. Shake the eggs gently and frequently during the treatment for uniform exposure of all eggs to acid solution. After washing, if water removal is incomplete in loose eggs, the HCl acid used for acid treatment becomes slightly diluted and hence after dipping, the temperature decreases consequently. Therefore, eggs should be dried properly before acid treatment. While for Chinese (oval cocoon) silkworm eggs, five minutes duration of dipping of eggs in hot HCl solution is adequate to stimulate embryonic development, the same is five and half minutes in case of Japanese (constricted cocoon) silkworm races. In case of hybrids, the duration of acid treatment is same as that of the female parent variety.

Cold Acid Treatment

It is also known as 'room temperature acid treatment' because hydrochloric acid used for the purpose is not preheated. In this case, concentration of acid is higher and duration of treatment is longer in comparison to hot acid treatment. But this method has certain advantages *viz.*:

- ☆ Simple and can be adopted without involving any special heating equipment/appliances.
- ☆ Conducted at room temperature and hence heating of acid is not required.
- ☆ It is safer than hot acid treatment, though it takes longer duration.
- ☆ Eggs do not drop-off the egg card.

- ☆ Unfertilized eggs crumple during treatment and hence can be identified very easily.
- ☆ It is more economical.
- ☆ As risk factor involve in this method is very minimum, it can be handled even by unskilled person.
- ☆ As no electricity for heating of acid is required, it can be practiced at all the places and at all the times.

Time of Treatment

Optimum time for cold acid treatment of bivoltine eggs is 15–20 hrs after egg laying preserved at 25°C.

Concentration of Hydrochloric Acid and Duration of Treatment

At room temperature (25°C), the optimum concentration of HCl acid solution in the range of 21–22 per cent with specific gravity of 1.108–1.110 is comparatively safe and dependable. If the specific gravity decreases below 1.10, the duration of treatment should be increased. Duration of treatment is decided according to the room temperature. If temperature of acid is lower than 25°C, the duration of treatment of eggs should be more. The duration of acid treatment at different temperature with acid concentration of 21 per cent (1.108 specific gravity) is detailed in Table 7.6.

Table 7.6: Acid Treatment Duration at Room Temperature

Oviposition Time	*Duration of Treatment (Minutes)/ Temperature of Solution*		
	24°C	*27°C*	*29°C*
10 hrs	60–70	40–70	40
15 hrs	60–80	60–80	40–50
20 hrs	60–100	60–80	40–50
25 hrs	60–100	60–80	40–50

Source: Narasimhanna, 1988.

Further, modified method of cold acid treatment can be practiced as indicated in Table 7.7.

Table 7.7: Ready Reckoner for Modified Cold Acid Treatment of Silkworm Eggs

HCl Specific Gravity at 25°C	*Acid Temperature (°C) and Dipping Duration (Minutes)*						
	20°C	*23°C*	*25°C*	*28°C*	*30°C*	*33°C*	*35°C*
1.090	—	100	90–100	70–100	60–100	40–80	30–60
1.100	90–100	60–100	60–100	40–80	30–60	20–40	20–30
1.110	40–100	30–60	30–50	20–40	20	10 -20	10

Removal of Acid and Drying

Soon after acid treatment, the eggs are washed in running water to remove all traces of acid from silkworm eggs. It is necessary to avoid irregular and poor hatching. Water temperature while washing eggs treated with HCl should be maintained within 15–30°C and water should continuously be stirred to remove hydrochloric acid as soon as possible. Blue litmus paper may be used to determine whether the acid is completely removed or not. After washing thoroughly, the eggs should be dried in well-ventilated room.

Cold Storage of Acid Treated Eggs

Silkworm eggs which are acid treated (hot or cold acid) for artificial hatching if incubated will normally develop and hatch in about 10 days. However, if due to some unavoidable circumstances, the hatching is to be delayed/postponed, the treated eggs are cold stored at 2.5°C up to 20 days without any harmful affect. If it is necessary to postpone the acid treatment itself, then eggs between the age groups of 20–24 hrs are first cold stored at 5°C for a week prior to acid treatment. In such cases, it is important that humidity during cold storage should be maintained at the optimum level of 75–80 per cent .

Acid Treatment After Cold Storage

This method can be applied in case hatching is required after 40 days of egg laying. In this method, 40–50 hrs old eggs on becoming dark brown after deposition at 25°C are cold stored at 5°C, for 35–70 days. Before subjecting the eggs for preservation, it is necessary to take following points into consideration:

- ☆ The preservation temperature after oviposition.

- ☆ Time or hours of cold storage.
- ☆ Time of acid treatment.
- ☆ Time specified for acid treatment should be followed strictly; otherwise it may lead to decrease in hatching.

The eggs are to be released from the cold storage to 24 ± 1°C through intermediate temperature of 15°C (2–3 hrs) on the desired day in between 35–70 days and are treated with acid to facilitate embryonic development. At this stage, cold storage inhibits diapause activity of eggs on the one hand and promotes hatching activity on the other. If cold storage period is shorter than 35 days, the hatching is affected considerably. Therefore, minimum period of cold storage of eggs at 5°C should be more than 35 days in order to obtain better hatchability. On the contrary, if eggs are cold stored for more than 90 days, they can hatch without acid treatment. The eggs, which are cold stored for more than 90 days should not be acid treated, as it is harmful and will result many dead eggs. Eggs after taken out from the cold storage are kept at room temperature for 1–2 hrs before dipping into hot or cold acid solution for acid treatment in order to stimulate artificial hatching. After acid treatment, eggs are washed properly in water and shade dried.

In this method, it is necessary that eggs must pass through intermediate temperature of about 15°C for 18 hrs at the time of removing them out of the cold storage or keeping them into cold storage from room temperature in order to avoid sudden change of the environmental conditions. Though the duration of immersing of eggs into acid in both the types *i.e.* acid treatment shortly after egg laying and that of cold storage is generally same, however, breed/hybrid dependency cannot be ignored.

Different types of acid treatment and their requirements are summarized in Table 7.8.

Assessment of Acid Treatment

Results of acid treatment can be assessed as indicated in Table 7.9.

Precautions during Acid Treatment

- ☆ Thermometer, hygrometer and timer should be perfect and to be tested before-hand to ensure that they are working properly.

Table 7.8: Acid Treatment and their Requirements

Particulars	*Hot Acid*	*Cold Acid*	*Acid Treatment for Chilled Eggs*
Specific gravity of HCl at 15°C	1.075	1.100	1.100
Conc. of HCl (%)	15	20	20
HCl temperature (°C)	46	25/27/29	47.8
Immersion duration (minutes)	Chinese: 4–5 Japanese: 5–6 European: 6–7	60–90 60–80 40–60	Chinese: 4–5 Japanese: 5–6 European: 6–7
Age of eggs (hrs)	20–24	15–20	a) 30–35 hr old eggs for 30–40 days b) 40–50 hr old eggs for 35–50 days
Colour of eggs	Yellow (before the appearance of brownish colour)	Yellow (before the appearance of brownish colour)	Yellow (before the appearance of brownish colour)

Table 7.9: Comparison of Properly, Insufficiently and Excessively Acid Treated Bivoltine Silkworm Eggs

Age of Eggs	*Properly Treated Eggs*	*In-sufficiently Treated Eggs*	*Excessively Treated Eggs*
24 hrs after treatment	Colour of eggs turns to pale or light brown. Colour change is slow. Appearance of slight depression on the upper surface of chorion.	Depression formation is not seen and eggs resembles to hibernated eggs	No colour change. Depression is very clear and deep.
2–3 days later	Clear formation of depression on the 2nd day and will be larger on the 3rd day	Very slow formation of depression. Even on the 3rd day, depression is not clearly seen.	Appearance of dead eggs with brownish colour.

☆ Purchase HCl from a reliable source only and check specific gravity (1.15 to 1.18) and colour of the acid (pale yellow).

☆ Only plastic/glass materials should be used in acid treatment room.

- ☆ Ensure proper ventilation in the treatment room as acid fumes in closed room will impair the health of workers in long-run.
- ☆ For personnel safety, it is suggested to wear protective glasses and apron.
- ☆ Check the water temperature while washing in running water. Too cold (less than 20°C) or too hot water should not be used.
- ☆ Solution of acid for treatment to be prepared preferably one day before acid treatment.
- ☆ Preferably two persons should be involved in acid treatment.

Biochemical Modulations Associated with Acid Treatment

Changes in Protein Synthesis Activity

One day (24 hrs) old bivoltine (pre-diapause) eggs are extremely sensitive to HCl treatment. All the eggs treated with HCl for specific duration starts development leading to good hatchability. On the other hand, three day old eggs when treated with HCl are found very less sensitive and hardly 10 per cent of eggs are noticed activated for development. This indicates that after 24 hrs of oviposition, certain changes in physiological processes concerned with diapause are progressed so much that they could not be reversed. As a result, the treatment turns out to be ineffective. The protein synthesis activity of the yolk cells is arrested when the eggs are treated within 24 hrs of oviposition with HCl. Acid treatment impedes embryonic role in the protein synthesizing system of the yolk cells which results in embryonic development like the non-diapausing embryos and the RNA synthesis of the yolk cells is not inhibited by acid treatment or cold storage or both.

Changes in Esterase–'A' Activity

Non-diapausing eggs after 20 hrs of oviposition, exhibit high esterase activity. On the other hand, diapausing eggs of the same age show very little activity. When acid treatment is performed on these eggs, the esterase-A activity rises abruptly approximately 30 minutes after HCl treatment. The activity becomes more and more pronounced in one or two hours after treatment almost matching with that of non-diapausing eggs of the same age group. This instant response of diapausing eggs to acid treatment suggests the possibility that the esterase-A which is contained in a bound form activate the inter esterase-A activity after treatment. In other words, acid treatment would cause the latent enzyme in diapausing eggs to set in motion. It is quite obvious that esterase-A activity is correlated with an active resumption of morphogenesis.

Changes in Amino Acids Activity

The concentration of most amino acids in treated eggs will not change during development but the level of arginine and phosphoethanolamine decreases quantitatively until the time of hatching. The fall in the level is very significant. The changes in glutamate, glutamine and praline in HCl treated eggs are very different from those in untreated eggs. The level of glutamate decreases slowly during development, while that of glutamine increases. The concentration of glutamine is increased by HCl treatment, but that of glutamate is not. Proline in the HCl treated eggs increases rapidly to a maximum by 5th day and then decreases rapidly; this change is quite unlike that in normal diapause eggs. Developmental changes in the concentrations of free amino acids in HCl treated eggs are similar to those in untreated eggs after resumption of embryogenesis. When embryogenesis proceeds in silkworm eggs *i.e.* after diapause of untreated eggs and in HCl treated eggs, the concentration of phosphoethanolamine decreases rapidly suggesting the importance of phospholipids formation for embryogenesis or the importance of phosphoethanolamine either as a precursor of other metabolic products or as an energy source.

Changes in Pigment Formation

Diapause eggs immediately after oviposition and till the germ band formation are yellow in colour. However, after a certain period of development, they gradually turn to brownish black or blackish brown due to the appearance of a specific pigment known as 'ommochrome'. This pigment spreads in the serosal cells and manifests the colour. Non-diapause eggs lack this pigment and hence the light yellow colour remains. Ommochrome is synthesized from trytophane. Tryptophane metabolism in insects has been studied extensively and the pathway from tryptophane to ommochrome has been reviewed in detail by Linzen (1974). In general, three major metabolites directly involved in the biosynthesis of the pigments are formyl-kynurenine, kynurenine and 3-hydroxyl kynurenine. The corresponding enzymes responsible for their production are kynurenine-formamidase, kynureninase and kynurenine–3 hydroxylase. Ommochrome is synthesized in developing ovaries and other tissues such as fat body during tryptophane metabolism as shown:

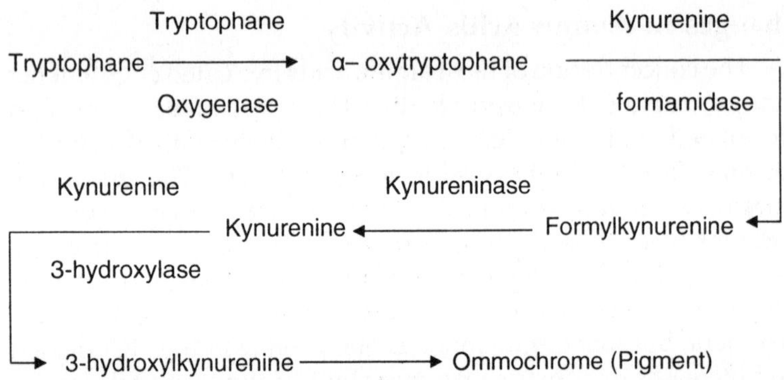

Changes in Glycogen Conversion

At the onset of diapause in silkworm eggs, glycogen is reported to decrease rapidly as it is converted into sorbitol and glycerol which are basically anti-freeze in nature (they increase the super-cooling ability, provide energy at diapause termination and stabilize the different enzymes even at low temperature). About 80 per cent of the glycogen is converted into sorbitol during the first 10 days after oviposition whereas glycerol gets accumulated during the later stages of diapause. At the termination of diapause, these polyols are converted into a glycogen store that can be utilized by the developing embryo as an energy source.

Changes in Glycogen Phosphorylase-A Activity

Glycogen phosphorylase-A is known to play a vital role in glycogen metabolism. It is highly probable that the key enzyme controlling polyol formation from glycogen at the initiation of diapause is phosphorylase-A. HCl treatment of diapause eggs after 20 hrs of oviposition blocks diapause initiation immediately and causes a peculiar temporal increase in phosphorylase-A activity.

Enzyme activity returns to its initial level in about 5 hrs and remains as that in non-diapause eggs without any further increase.

Changes in Lipid Metabolism

In diapause eggs, the lipid concentration in egg contents is about 60–70 mg/g of eggs from 2 hrs to 40 days after oviposition. On the contrary, in the HCl treated eggs, the lipid concentration decreases (68–32 mg/g of eggs) after oviposition until hatching. It is interesting to note that the lipid concentration in egg shell does not change significantly during diapause stage as in HCl treated eggs.

Hot acid treatment as well as cold acid treatment if carried out properly at appropriate time, duration and age have been found equally efficient and are bound to give excellent results. However, the ultimate choice is left to the user who should take a final decision considering:

- ☆ The type of eggs to be treated (pure univoltine or bivoltine or hybrids)
- ☆ Age of eggs at the time of acid treatment
- ☆ Seasons of acid treatment
- ☆ Quantity of eggs to be treated
- ☆ The time at disposal
- ☆ Availability of infrastructure and other facilities for acid treatment

Incubation

Quality silkworm seed is the basic requirement of the sericulture industry. The crop performance and silk productivity are directly related to the quality of seed. It is very difficult to define the term 'Quality of silkworm seed' but for all practical purposes, the quality silkworm seed can be considered as eggs produced from good cocoons under optimal environmental conditions. Quality can deteriorate due to improper handling during its processing, preservation, hibernation, acid treatment, chilling, incubation, transportation etc. Incubation is one of the most important techniques in seed handling. Incubation is an acclaimed important step to obtain uniform hatching, normal growth of larvae, quality of cocoon crop and quality of cocoons. Therefore, the term incubation can be defined as preservation of silkworm eggs under optimum environmental conditions of temperature (25 ± 2°C), RH (75 ± 5 per cent) and photoperiod (16 L: 8 D) during embryonic development, so as to suit the silkworm eggs to develop normally to provide uniform and maximum hatchability on the expected day. The term incubation is used especially to those eggs, which are naturally active or artificially activated and not to those eggs, which are kept under aestivation. Thus, incubation of silkworm eggs is conducted under manipulated environmental conditions, which enables them to develop normally.

Objectives of Incubation

Incubation aims at:

1. To promote proper development ensuring that eggs of different breeds/hybrids should hatch on the desired date and time and it should be above 90 per cent on a single day.
2. To enforce uniform hatching to produce healthy larvae and stable cocoon crop.
3. To maintain the voltinism of the breed.
4. To stabilize the hibernating characters and potency value of the bivoltine eggs.
5. To maintain uniform larval growth for successful cocoon crop.

Types of Incubation

Incubation is of three types:

Interruption Incubation

In the interruption method, the temperature at the beginning of incubation is kept at 15–18°C and after one or two days the incubation is allowed for four days at 22–23°C up to blastokinesis stage. At the final stage of incubation, the temperature is maintained at 25°C until hatching.

Progressive Incubation

In the progressive incubation method, the temperature in the beginning is maintained at lower level and thereafter it is kept constant around 25°C until hatching.

Kyuri Incubation

This method is not in practice now-a-days as the technique of artificial hatching method is well adopted. However, in this method the temperature in the beginning is maintained at 20–23°C and during the development of embryo when thoracic appendages appear or just before the blastokinesis, the temperature is brought down to 15°C. This condition is continued until just before hatching. At the time of hatching once again, the temperature is increased to 24–25°C. This incubation method is adopted when it is necessary to obtain non-diapause eggs.

Methods of Incubation

Incubation is carried out generally by two methods:

1. Constant temperature incubation
2. Raised temperature incubation

In constant temperature incubation method, incubation is carried out at a constant temperature of 25–26°C from beginning to till hatching. This method is practiced for acid treated and non-hibernated eggs. Raised temperature incubation method is used for incubation of hibernating eggs. The hibernating eggs soon after release from cold storages are preserved at 10–15°C for 3 days, 18–20°C for 2 days; 23–24°C for 4 days and 25–26°C till hatching. This avoids sudden shock of temperature to delicate silkworm eggs and helps in uniform development of embryos and hatching. However, generally, constant temperature incubation is in practice.

Devices for Silkworm Egg Incubation

Incubation can be practiced by adopting the following techniques:

Room Incubation

In places where cold storages are available, exclusive incubation rooms are constructed. These rooms have provision of maintaining optimum temperature and humidity conditions, provision of regulating the photoperiod besides good ventilation. Incubation rooms should not be overcrowded with silkworm eggs as it increases the carbon-dioxide content during development.

Mud Pot Incubation

Mud pot incubation is ideal for individual farmers for incubation of 250 to 300 Dfls. The well baked earthen pot with a wide mouth of 15″ diameter is chosen. The mud pot is filled with a clean sand layer of 2–3″ thick and made wet depending on the need and requirement. It is provided with a row of holes of about half an inch diameter situated 2–3″ above the sand level. The pot is covered with wet muslin cloth. The layings are hanged inside the pot with the help of a rod and the pot is placed over an ant well. Mud pot incubation can be utilized whenever temperature prevails higher than optimum. It is more suitable method of incubation in hot and arid (dry) areas.

Double Brick Walled Incubation Chamber

This method of incubation is suitable for individual grainages, CRCs and big sericulture farmers. The chamber is constructed using brunt bricks, sand and cement. Gap of 3″ is maintained between outer and inner walls. The gap is filled with loose clean sand. The standard size of outer wall is 6′ x 4′ x 3′ and inner wall 4′ x 2′ x 3′ with a gap of 3″ between them. This incubation device can accommodate 5000 – 6000 Dfls. The egg sheets are hanged in the chamber with sufficient gap between them. The sand layer is soaked periodically with water depending upon the prevailing environmental conditions during incubation. The temperature and humidity can be maintained to near optimum conditions by this method. This method of incubation is suitable during summer months.

Tray Incubation

The layings can be incubated in clean rearing tray by providing paraffin paper and wet foam pads. During favourable seasons, the silkworm egg sheets are spread in the rearing trays for incubation. The rooms should be heated with the help of electric heater up to the required temperature. Boiling water in wide mouthed vessel on charcoal stove can increase humidity. This helps in increasing temperature and humidity both.

Incubation in Incubator

In India, to maintain the required temperature and humidity for incubation, the silkworm eggs are incubated in BOD incubator.

Incubators with light facility can also be used for incubation of silkworm eggs. The temperature in incubator is set to the required level and humidity is maintained by keeping a tray of water inside the incubator preferably at the bottom.

Factors Associated with Incubation

The main factors related to incubation are disinfections, transportation, temperature, relative humidity, photoperiod, aeration and black boxing, which determine the various activities of life.

Disinfection

Rooms and equipments where incubation has to be carried out should be thoroughly disinfected prior to commencement of incubation. Silkworm eggs should be surface sterilized with 2 per cent formalin for 10–15 minutes before and after cold storage of eggs.

Transportation

Irregularities in hatching and subsequent crop losses can also be attributed to improper transportation methods. Extreme environmental conditions such as high temperature coupled with low humidity and oxygen deficiency during the course of transportation are harmful. Therefore, care should be taken to provide optimum incubation conditions during transportation. Efforts (Rajan and Himanthraj, 2005) should be made to avoid:

- ☆ Exposure of eggs to direct sun light and high temperature.
- ☆ Transportation of eggs in sealed boxes without perforations.
- ☆ Storing of eggs in enclosed places with poor ventilation.
- ☆ Direct or indirect contact of eggs with tobacco.
- ☆ Contamination of eggs with pathogens.
- ☆ Exposure of eggs to fumes, fertilizers, petroleum and insecticides.
- ☆ Physical shocks to eggs.

It is advisable to transport silkworm eggs during early developmental stages and at the time when temperature fluctuation is less *i.e.* during morning or evening time specially by keeping them in transportation boxes fabricated for the purpose.

Temperature

Temperature during incubation is one of the most important factors that play a vital role and decides the growth and development of embryos, uniformity of hatching, hatching per cent, vigour of larvae and quality of cocoons besides voltinism. Although the embryonic development takes place between 10–30°C but their development is not uniform leading to irregular hatching over a period of time. The optimum temperature of 25 ± 1°C is ideal for incubation at which the embryonic growth is normal. Any fluctuation from the optimum during incubation affects the development of embryos. The relation between incubation temperature and time required for hatching is presented in Table 7.10.

Table 7.10: Incubation Temperature and Time Required for Hatching

Sl.No.	Incubation Temperature (°C)	Incubation Duration (Days)
1.	10	70-80
2.	15	30-33
3.	20	15-16
4.	25	9-10
5.	30	8-9

Temperature above 30°C is highly deleterious as hatching is significantly affected, and at 33°C hatching is almost negligible. Temperature during incubation also affects voltinism as the embryonic stages are most sensitive to it. Bivoltine eggs incubated at 25°C and above produce moths that lay hibernated eggs, while those incubated at temperature below 25°C produce moths that lay mixed and fully non-hibernated eggs (20°C mixed and 15°C non-hibernated eggs). Silkworm eggs during course of its development under optimal conditions show approximately 12 per cent weight loss. This loss is due to the oxidation of nutrients required for the growth of the embryo during which metabolic water, CO_2 and energy are released (Reddy *et al.*, 2005). Temperature above 30°C results in abnormal weight loss which leads to desiccation of eggs affecting hatching per cent. Influence of order of egg maturation on weight of eggs, glycogen and water contents (moisture) in newly laid eggs of silkworm is presented by Reddy *et al.*, 2005 (Table 7.11). It is established that embryos can tolerate high temperature during the first two days of incubation, but higher temperature for more than two days killed embryos.

Table 7.11: Influence of Order of Eggs Maturation in Bivoltine Silkworm Breeds of NB_4D_2

Sl.No.	*Time of Oviposition*	*No. of Eggs/gm*	*Weight of Eggs (mg)*	*Moisture %*	*Glycogen (mg/gm Wet Weight)*
1.	4 pm–5 pm	1670	0.59	63.40	20.93
2.	5 pm–6 pm	1750	0.57	62.86	23.18
3.	6 pm–10 pm	1750	0.57	62.79	33.15
4.	10 pm–10 am	1820	0.55	61.52	31.91
5.	10 am–10 pm (2nd day)	1820	0.55	61.21	32.68
	Average	1762	0.57	62.36	23.37

Humidity

Humidity is an important factor, which directly affects the physiology of silkworm eggs. Too dry or too wet air is harmful to the physiology of developing embryos. The humidity range of 70–80 per cent is ideal for maintaining the normal development of silkworm eggs. Humidity less than 60 per cent results in loss of water from silkworm eggs and humidity of 90 per cent and above leads to retention of physiological waste water resulting in poisoning of embryos. Humidity in combination with temperature has very strong effect on the developmental physiology of eggs. Though a temperature of 25 ± 1°C is ideal for incubation but if humidity is as low as 30 ± 5 per cent, the ideal temperature itself becomes dangerous for the development of eggs. It not only affects hatching but also retards growth and prolongs the incubation period. Low humidity during pinhead and blue egg stage makes the gluey substance that covers the eggs rather hard as cement and the larvae find it difficult to bite the eggshell and eventually fail to hatch. A combination of high temperature and low humidity, results in desiccation and majority of eggs fail to hatch. The weight of newly hatched larvae decreases significantly when incubated at low humidity and high temperature due to desiccation and improper utilization of yolk. If humidity increases to 90–100 per cent during incubation, weak trimoulter larvae silkworm appears during rearing.

Table 7.12: Recommended Temperature and Humidity During Incubation of Eggs

Type of Eggs	*Temperature*	*Duration*	*Humidity*
Acid treated eggs	24–26° C	Up to hatching	75 ± 5 per cent
Hibernated eggs	15° C	3 days	
	24–26° C	Up to hatching	

Photoperiod

Photoperiod has a direct effect on hatching and voltinism during incubation period in the bivoltine silkworms. From the very beginning of incubation up to the eye spot stage, light hastens the embryonic development. However, during final hatching stage, darkness inhibits development. Bivoltine silkworm eggs should be incubated at 16 hrs lights and 8 hrs dark to maintain the diapausing nature in the ensuing generation. The intensity of light on egg surface should be above 10 lux. In case of low temperature incubation, the eggs are kept in darkness throughout the day or the light conditions per day should not exceed 12 hrs. If eggs are allowed to remain under light conditions throughout the day, the hatching will not be uniform. The most critical stage of the embryo, which is sensitive to photoperiod, is from the stage of appearance of thoracic appendages. In order to obtain maximum hatching within a short period, the eggs in eye spot stage are kept in darkness and on the day of hatching, they are suddenly brought out to bright light conditions.

Aeration

Since the eggs undergoing incubation respire vigorously, the room should be airy, so that there is no hindrance to free respiration. Aeration from the middle of the incubation plays very important role. Good ventilation provides good circulation of air, which helps in driving out poisonous gases produced due to metabolic activity of silkworm eggs. It is established that carbon dioxide level above 0.3 per cent is deleterious and kills the developing eggs. Therefore, incubation room should be properly ventilated or aerated.

Black Boxing of Eggs

Black boxing technique provides synchronized hatching on a single day and within in a short period. The eggs, which are not black boxed, hatches irregularly over a period of time. In order to obtain synchronized hatching in a short period, photoperiod is manipulated during the later stage of egg development especially at the time when egg development reaches to pinhead stage (head pigmentation stage), the eggs are subjected for complete darkness known as 'scotophase' and exposed to light after 48 hrs. This procedure is based on the fact that faster egg development takes place in the light up to head pigmentation stage and darkness enhances uniform development from pinhead stage up to onset of hatching. This technique of providing complete darkness to developing embryos from pinhead stage to the onset of hatching is known as 'black boxing technique'. It has two advantages:

- ☆ To force late maturing or lagging embryos to accelerate their development and join there developed counterparts.
- ☆ The insects being photosensitive in nature, darkness inhibits hatching process largely. This inhibition is intricately related to the 'biological clock' operating in the nervous system.

Methods of Black Boxing

The methods of black boxing are:

Dark Room Method

The entire room where silkworm eggs are incubated can be made dark by switching off the lights when eggs reached to head pigmentation stage and exposed to light only after 48 hrs.

Incubator Method

The layings are incubated in total darkness in incubators when they reach head pigmentation stage and exposed to light only after 48 hrs.

Black Cloth Method

The layings kept in trays for incubation are covered with double-layered black cloth when developing embryos reaches head pigmentation stage and exposed to light after 48 hrs.

Wooden Box Method

A wooden box painted on inner side with black paint can be effectively used for incubation. Foam pads or wet cloths or wet paper can be kept on the floor of the box to increase the humidity if required. Eggs after reaching to head pigmentation stage are kept in these boxes and exposed to light after 48 hrs.

Black Paper Method

The layings are kept in black paper cover, which are sealed by using clips to avoid entry of light. The clips are opened and eggs are exposed to light after 48 hrs on the expected day of hatching.

Stage of Black Boxing

Silkworm eggs should be black boxed when majority of eggs (80 per cent) has reached to head pigmentation stage. Black boxing of eggs at body pigmentation stage (blue egg stage) is not ideal, as light stimulus would have already been perceived by the brain cells of embryos, as a result hatching will be initiated inside the black box itself (Rao *et al.*, 1998). The head pigmentation stage is identified by a small black dot in the eggs near the micropylar end. The black spot appears mostly on the 7th day in multivoltine breeds and on the 8th day in bivoltine breeds under optimum incubation conditions.

Duration of Black Boxing

The period of black boxing depends on the stage of embryo prior to black boxing. In multivoltines, the pigmented stage can easily be identified; as such 48 hrs black boxing is sufficient for synchronized developments and hatching. However, in bivoltines, the change of colour of egg to pin-head stage is not clearly visible due to the pigmented serosa and requires expertise in identifying the stage. The appearance of head pigmentation and consumption of serosa are simultaneous process. After complete consumption of serosa, the pin head becomes clearly visible. Thus eggs can be black boxed for 72 hrs after the appearance of head pigmentation stage. However, if uniform head pigmentation in majority of eggs are observed, they can be black boxed for 48 hrs and then exposed to light.

Light Exposure to Stimulate Hatching

A photoperiod of 16 hrs light and 8 hrs darkness during incubation up to pin-head stage and then shifting to darkness and later sudden exposure to bright illumination enable uniform hatching. The hatching date can be determined by the embryonic development with reference to the colour of the egg and distinctness of the egg dimple. Hatching occurs within 2–3 hrs after exposure to light. The layings from the dark phase are exposed to bright light at about 6 A.M. In the absence of electricity, the layings can be brought under natural light. A relative humidity of 70 per cent is ideal at the time of hatching.

Effect of Incubation Environment on Voltinism

The major environmental factors affecting voltinism are:

Temperature

Temperature during incubation has greatest influence on voltinism. In other words, temperature during incubation plays an important role in determining that whether bivoltine silkworms will lay hibernating or non-hibernating eggs. When bivoltine eggs are incubated at a temperature of 25°C or above, the resulting moths will lay hibernating eggs, and when incubated at a low temperature of 15°C, they will lay non-hibernating eggs regardless of environmental conditions during larval and pupal stages. An intermediate temperature of 20°C during incubation will result in production of mixture of hibernating and non-hibernating eggs (Table 7.13).

Photoperiod

Light during incubation also influences voltinism but the effect is not evident if temperature is above 25°C or below 15°C. At the intermediate temperature, if long photoperiod is provided to the developing embryos, the resulting moths will lay higher rate of hibernating eggs. On the contrast, the short photoperiod induces more non-hibernating eggs. It is established that the influence of light is next to temperature.

Humidity

The influence of humidity on voltinism is next to light. Humidity does not affect voltinism if incubation temperature is above 25°C or below 15°C. At intermediate incubation temperature, high humidity is favourable for inducing more hibernating eggs, while low humidity favours non-hibernating eggs. Temperature, light and humidity affect the voltinism of silkworms at the stage between when embryo begins to form the thoracic legs and when the head begins to be pigmented. The basic requirement for the bivoltine to lay hibernating eggs is a temperature above 25°C. If eggs are incubated at a temperature below 15°C and kept in dark and dry condition, most of the bivoltine races may produce non-hibernating eggs in the resulting moths.

Table 7.13: Influence of Temperature During Incubation on Various Stages of Development of Bivoltine Silkworm

Incubation Stage	*Grown Larval Stage*	*Mounting/ Pupal Stage*	*Voltinism of Eggs Laid*
High temperature (Above 25°C)	High temperature (Above 25°C)	High temperature (Above 25°C)	Hibernating
	Low temperature (18°C)	Low temperature (20°C)	Non-hibernating
Low temperature (15°C)	High temperature (Above 25°C)	High temperature (Above 28°C)	Non-hibernating
	Low temperature (18°°C)	Low temperature (20°C)	Non-hibernating
Intermediate temperature (Above 20°C)	High temperature (Above 25°C)	High temperature (Above 28°C)	Mostly Non-hibernating
	Low temperature (18°C)	Low temperature (20°C)	Mostly Non-hibernating

Cold Storage of Blue Eggs and New Born Larvae

Hatching of eggs during incubation can be delayed by cold storing the eggs at blue egg stage at 5°C for 3–5 days. Newly hatched larvae if necessary can be cold stored for 3 days at 7–10°C. The humidity in the cold storage room should be maintained at 75–80 per cent .

Rotten Eggs Before Head Pigmentation

Sometimes rotten eggs such as brown rot, red rot and grey rot eggs are observed during incubation period. Several factors are responsible for such type of damage of eggs:

- Exposure of eggs to high temperature (> 28°C) and high humidity (> 90 per cent) during oviposition and incubation.
- Storage of hibernating eggs for too long period under dry and high temperature conditions.
- Over stimulation of hibernating eggs during acid treatment.
- Contact of eggs by pesticides, nicotine, mosquito repellant, incense, oil or sticky material such as glue or gum.

Dead Eggs After Head Pigmentation

During incubation, two types of embryo death can be observed *i.e.* one at head pigmentation stage and the other when eggs ready to turn bluish. In these cases, embryo is normally formed but the death occurs before hatching due to:

- Too high temperature during incubation (> 28°C).
- Too low relative humidity during incubation (< 50 per cent). This condition while causing death of the embryo, also results in stray hatching (un-uniform hatching) and under weight of newly born larvae.
- Contact of eggs with pesticides, nicotine, mosquito repellant, incense or other toxic materials.

Quality Check

Quality check is necessary at every stage of activity *i.e.* at in-coming, in-process and also at product stages in order to produce good quality of eggs. The check points are:

- Minimum two times disinfections of seed rearers' houses.
- Supply of disease free layings (Dfls) meeting fecundity norms, incubation, black boxing, brushing, larval examination and crop supervision.
- Purchase norms and procedures, conditions of seed cocoons received, its pupation rate, uniformity, shape, size and purity of breeds etc.
- Conditions of cocoons preserved, conditions of pupae after sex separation, temperature and humidity maintenance.

- ☆ Conditions of moth emergence rate, healthiness, activeness and disease freeness.
- ☆ Conditions of preservation of male moths in cold storages, pairing duration etc.
- ☆ Conditions of production of eggs, recovery and unfertilized eggs, oviposition room ambience.
- ☆ Conditions of handling of eggs, non-hibernating eggs, egg colour and dead eggs besides hatching percentage etc.
- ☆ Fujiwara's method of mother moth testing.
- ☆ Specific gravity and temperature of HCl for acid treatment.
- ☆ Conditions of egg supplied and disease freeness.
- ☆ The quality standards for production of silkworm seed shall not be below the recommended level.

A good seed producer looks for internal checks and balances and ensure that seed certified for marketing has passed all tests and therefore likely to perform well. It is to be ensured that in silkworm seed production, stipulated benchmark is met fully. All these measures would help to ensure that quality silkworm seed produced is of highest order and seed provide enough strength to the commercial silkworm seed producers to exploit the potential.

Chapter 8

ADAPTATIONS

The most important factors in environment that influence the physiology of insects are temperature and humidity. Insects display a remarkable range of adaptations to changing environments and maintain their internal temperature (thermoregulation) and water content within tolerable limits, despite wide fluctuations in their surroundings. Adaptation is a complex and dynamic state that widely differs in species. Surviving under changing environment in insects depends on dispersal, habitat selection, habitat modification, relationship with ice and water, resistance to cold, diapause and developmental rate, sensitivity to environmental signals and syntheses of variety of cryoprotectant molecules. The mulberry silkworm (*Bombyx mori*) is very delicate and sensitive to environmental fluctuations and unable to survive naturally because of their domestication since ancient times. Thus, the adaptability to environmental conditions in the silkworm is quite different from those of wild insects. Temperature, humidity, air circulation, gases and photoperiod etc. shows a significant interaction in their effect on the physiology of silkworm depending upon the combination of factors and developmental stage affecting growth, development, productivity and quality of silk.

Adaptation is characteristic of an organism that makes the adjustment of living matter to environmental conditions either in its lifetime (physiological adaptation) or in a population over many generations (evolutionary adaptation). The ability to adopt is a fundamental property of life and constitutes a basic difference between living and non-living organism. Insects display a remarkable range of adaptations to changing environments. The extensive literature on insect seasonality reveals great diversity and complexity in the adaptations that withstand seasonal adversity and synchronize development with the seasons (Danks, 2007). The basic structure of these adaptations lies in the specific nature of the environment, the components of which in insects are reviewed, especially from the viewpoint of aspects to understand the structure and functions of adaptations. The component responses include dispersal (Bale, 2002; Brower, 1995; Genkai, *et al.*, 2005), habitat selection and habitat modification (Powell *et al.*, 2006), resistance to cold (Ramlov, 2000), dryness (Dautel, 1999) and food limitation (Anderson, 1974), diapause (Ando, 1974; Nakamura and Numata, 2000), modification of developmental rate (Honek and Kocourek, 1990), sensitivity to environmental signals (Danks, 2005), life-cycle patterns (Danks, 2000), variation in phenology (Chen *et al.*, 2003) and development (Bennett *et al.*, 2005; Danks, 2007) and all these are the result of environmental pressures. Many types of adversity can prevent the activity or threaten the survival of insects *viz.* adaptations in response to anoxia (Hoback and Stanley, 2001; Hodkinson and Bird, 2004), as well as to ice scour, spring spates, floods and other extremes or disturbances (Lytle, 2002; Prowse and Culp, 2003). Insects survive the cold of winter by a wide array of adaptations that have been reviewed many times (Bale, 2002; Danks, 2005; Duman *et al*, 1991). These studies have led to the discovery of more and more elements that contribute to insect cold-hardiness. The critical element of adaptations to changing environment is how insect respond to the presence or reliability of the environmental signals, not just the environmental conditions (Danks, 2006). Insects use one or combination of photoperiod, temperature, thermoperiod, moisture, food and other factors to assess the current or future suitability for their development and or reproduction. Since silkworms are poikilothermic animal, environmental conditions in general, temperature and humidity in particular plays a decisive role in the determination of most of the physiological processes.

I. Temperature

Among various environmental elements, temperature is most critical factor, associated closely to climates and effective through lower and upper limits and through heat accumulation. Lower developmental limits for insects average about 10°C or 11°C (Honek and Kocourek, 1990) and in most temperate species upper limits lie between 20°C and 35°C (Danks, 2007). The impact of temperature is modified by habitat and other physical conditions. Many types of adversity can prevent the activity or even survival of insects. Adaptations are known in response to anoxia (Hoback and Stanley, 2001; Hodkinson and Bird, 2004) as well as to spring spates and other extremes or disturbances (Lytle, 2002; Prowse and Culp, 2003). However, abundant information is available on survival of insects in cold, dryness and lack of food. The adaptation of insects to cold hardiness has been reviewed by various workers (Bale, 2002; Danks, 2005). These studies have led to the discovery of more and more elements that contribute to insect cold hardiness and the same are summarized in Table 8.1.

Table 8.1: Elements of Insect Cold Hardiness (*c.f.* Danks, 2006)

Elements	*Sample Features*	*Main Effects on Cold Hardiness*
Region	Climate, weather	Temperature and its variation, freezing rate.
Microhabitat	Choice, site features, modification	Insulation, protection against inoculative freezing, freezing rate, effects of ice.
Crystallization	(i) Freezing resistance, freezing tolerance	Absence or presence of ice.
	(ii) External nucleators, internal nucleators, ice nucleating proteins	Ice formation.
Water relations	Status, availability, dehydration	Ice amount, un-freezable water, water management and solute interactions, super cooling by dehydration.
Cryo-protectants	Solutes, antifreeze proteins, shock proteins	Inhibition or modification of ice or injury.
Other molecules	Enzymes, antioxidants	Modification of functions or injury.

Insects survive the cold of winter by a wide range of adaptations that have been reviewed many times and elements of these responses are compiled, summarized and presented in Table 8.2. Insects can be injured at low temperatures of 0°C (chilling injury) (Turnock and Fields, 2005). Injury from cold appears to be associated with a breakdown of membrane structure and membrane based ion gradients (Zachariassen *et al.*, 2004); compromised protein structure and enzyme function. Adaptation to temperature below 0°C, include modification in the saturation of membrane fatty acids to maintain function at low temperatures. Shock proteins (Craig *et al.*, 1993) produced in response to chilling are also presumed to protect against low temperatures (Denlinger, 1991; Relina and Gulevsky, 2003). At temperatures below 0°C, most species remain unfrozen because they supercool. Super cooling is enhanced by solutes of low molecular weight (Lee, 1991; Ramlov, 2000). Many species have more than one kind of solute *viz.*, both glycerol and sorbitol. The freezing process depends on the supply of water molecules 'bound water' (Block, 2002; Block and Zettel, 2003). The cold hardiness of several species in extreme environments includes marked dehydration (Bennett *et al.*, 2005; Ring and Danks, 1994). Information is also available on the pathways of association of other elements in cold hardiness, such as cholesterol (Yi and Lee, 2005), membrane transport of cryoprotectants and water (Izumi *et al.*, 2006). These observations indicate that cold hardiness is not a static condition instituted for the winter, but a series of complex developmental and metabolic patterns. The effects of cold across different temperature ranges and with different times of exposure interact for both chilling and freezing injury, and there is some repair of cold injury during winter (Renault *et al.*, 2004; Turnock and Fields, 2005). Diapause is linked to cold hardiness to different degrees. The mechanism of relationship between cold hardiness and diapause is very complex. In several species some aspects of the diapause programme are prerequisite for some aspects of cold hardiness (Danks, 2005). Susceptibility to cold shock may also depend on the diapause status (Pitts and Wall, 2006).

Among the various environmental factors that influence the silkworm cocoon crops, the most important are temperature followed by humidity. Since silkworms are cold-blooded animals, temperature has a direct effect on the growth, development and physiological activity, nutrient absorption, digestion, blood circulation, respiration etc. Temperature plays an important role in egg hatching, larval growth and quality of the cocoons produced, eclosion, adult fertility

Table 8.2: Common Mechanism of Direct Resistance to Cold in Insects (*c.f.* Danks, 2007)

Mechanism	*Sample Systems or Substances*
	Survive chilling
(*a*) Protect membranes or membrane functions	(a) Membrane transition temperature lowered by changing composition of fatty acids.
(*b*) Protect proteins, enzymes and other functions	(b) Potential roles of shock proteins and other molecules.
	Prevent freezing
(*a*) Lower haemolymph melting point	(*a*) Manufacture of low molecular weight solutes including glycerol, other polyhydric alcohols and sugars.
(*b*) Lower haemolymph freezing point	(*b*) Antifreeze proteins lower the freezing point (this action is enhanced by glycerol and other molecules).
(*c*) Reduce nucleation sites	(*c*) Adaptation in structural proteins etc.
(*d*) Eliminate nucleators	(*d*) Empty gut.
(*e*) Mask nucleators	(*e*) Nucleation sites masked by nucleation inhibitors.
(*f*) Reduce water available for freezing process	(*f*) Associate water with cell constituents; dehydrate passively by readily losing water through cuticle.
	Survive freezing
(*a*) Limit super cooling to reduce impact of freezing events	(*a*) Manufacture of nucleating proteins, cuticular structure readily allowing inoculation.

Contd...

Table 8.2–Contd...

Mechanism	*Sample Systems or Substances*
(*b*) Reduce amount and rate of formation of ice	(*b*) Manufacture low-molecular weight cryoprotectants.
(*c*) Protect membrane	(*c*) Protection provided by association of various cryoprotectants with membranes.
(*d*) Protect other functions	(*d*) Protection provided by association of various cryoprotectants with membranes.
(*e*) Rapid cold hardening	(*e*) Enhanced resistance caused by brief cooling beforehand.
(*f*) Prevent recrystallization	(*f*) Manufacture of antifreeze proteins, manufacture of specific recrystallization inhibitors.
(*g*) Modify water status	(*g*) Un-reactive glassy states mediated especially by carbohydrates such as trehalose.
Dynamic adjustment	
(*a*) Seasonal adjustments of cryoprotectants/ Rapid cold hardening	(*a*) Solutes and antifreeze proteins change through the winter in some species, changing super-cooling point and other features.
(*b*) Freezing induced change in super-cooling point	(*b*) Freezing exposure lower super-cooling points in some species.
(*c*) Linkages with diapause	(*c*) Diapause is prerequisite for some elements of cold hardiness in some species.
(*d*) Seasonal adjustment of mitochondria	(*d*) Mitochondria reduced during winter cold and inactivity.
(*e*) Ongoing repairs	(*e*) Warmer intervals permit injury caused by cold exposures to be repaired in some species.
(*f*) Interaction between antifreeze proteins	(*f*) Enhance action.

etc. With the increase in temperature, the larval growth and development is accelerated resulting reduction in larval duration, cocoons of lower weight and quality, while at low temperature growth and development is slow leading to prolong larval period, abnormal growth and sensitivity against several diseases. The optimum temperature for healthy growth and higher production of quality cocoon is 26–28°C for 1st instar and 2nd instar, 24–26°C from 3rd to 5th instars and 22–23°C for mounting and 24–25°C for egg incubation. However, for technical reasons, temperature of 25°C–26°C is used for all stages. In case of multivoltine breeds slightly higher rearing temperature than that required for bivoltine is necessary. The room temperature is generally low during winter and high during summer months and therefore suitable measures are adopted to regulate it to the optimum level. Rearing of 4th and 5th instar larvae at low temperature is better for converting larger percentage of food into silk, since food stays for longer period in the stomach and the digestion and absorption of nutrient is better. Even though, low temperature has certain advantages, but must be avoided as they decrease the growth rate and increase rearing period affecting the cost of rearing (Upadhyay and Mishra, 1994) besides inviting various diseases.

Temperature and Embryonic Development

The effect of temperature on the development of silkworm has been studied extensively but much attention has not been paid on the effect of temperature on embryonic development. It has been reported that in exothermic organisms, when rate of development is plotted against temperature, a sigmoidal curve is obtained with an almost linear correlation in central temperature range. Attempts have been made to describe the curve more accurately using different mathematical formulas (Bottrell, 1975; Laudien, 1973) but there are also other variables which may affect the rate of embryonic development, such as adaptation phenomenon, genetic variation besides health and age of the organism. Kalthoff (1971) reported that in a given species, different developmental phases may be characterized by different upper and lower limits of temperature and inferred that the correlation between temperature and rate of development may not be the same for all stages of embryonic development. Under extreme environmental conditions, particularly in climates with very hot or cold seasons, a resting stage called 'diapause' (Denlinger, 1985) may ensure the survival of the species. This is important for early embryonic stages, which are often more sensitive to extreme temperatures than any other stages in the life cycle of an insect. However, in some species, sub-zero temperature cannot be avoided and physiological adaptations are required in such cases. While some insects have become freezing tolerance, others have developed mechanism to keep their body below the freezing point, a phenomenon referred as 'super cooling'. The super cooling is common in insect eggs. Eggs of lepidopteron *Zeiraphera diniana* may supercool to minus 51.3°C (Bakke, 1969). Salt (1959) stated that the depression of the freezing point and increased super cooling capacity can be attributed to the accumulation of glycerol in many species. In *Bombyx mori* eggs, at the initiation of diapause, large amounts of sorbitol and glycerol are synthesized from previously accumulated glycogen. After diapause, the reaction is reversed and glycogen is synthesized (Yaginuma and Yamashita, 1979).

$$\text{Glycogen Sorbitol} \longleftrightarrow \text{Glycerin}$$

With the initiation of diapause, the above reaction is right oriented while during termination; it is left oriented (Yaginuma and Yamashita, 1977 & 1978). The experiment with 14^C glycine showed that sorbitol is totally derived from glycogen, while glycerin is produced only when glycogen content reached the lowest level. About the mechanism of reversible reaction, following steps exists:

$$\text{Glucose} + \text{NADPH} \longleftrightarrow \text{Sorbitol} + \text{NADP}^+$$

$$\text{Glucose-6-p-fructose-6-p} + \text{NADPH} \longleftrightarrow \text{Sorbitol-6-p} + \text{NADP}^+$$

$$\text{Glyceraldehydes} + \text{NADPH} \longleftrightarrow \text{Glycerin} + \text{NADP}^+$$

$$\text{2–OH acetone-p} + \text{NADH} \longleftrightarrow \text{2-p glycerin} + \text{NAD}^+$$

Diapausing silkworm eggs chilled at 5°C for at least three months when transferred to 25°C, the eggs resumed embryogenesis and hatches within two weeks (Yamashita *et al.*, 1988). Soaking the pre-chilled diapausing eggs into hot solution can also bring about hatching (Yamashita, 1984). In these eggs, sorbitol content began to decrease after continuous chilling for at least two months or by treatment with HCl on one month chilled eggs. Greater fall in sorbitol content took place in eggs chilled for longer periods. Conversion of sorbitol to glycogen does not take place if eggs are kept continuously at 25°C (Yamashita *et al.*, 1988) even for more than six months. After diapause, all sorbitol in the eggs is converted into glycogen. During embryogenesis, the main carbohydrate consumed is glycogen. The glycogen-sorbitol-glycogen metabolic process is roughly the same as the process of diapause onset, maintenance and termination. In other words, the change of carbohydrate metabolism of the silkworm eggs is a close relative to the phenomenon of diapause.

Temperature and Physiological Effects

Temperature is a parameter in developmental cycle which can be manipulated experimentally but their effect is very complex for interpretation. The physiological explanation for embryonic death after exposure to lethal temperature is likely to be highly complex and probably species specific. Kittlans (1961) stated that temperature above 33°C leads to abnormal development in one species of *Leptinotarsa* and two species in *Epilachna* and suggested that at these temperatures yolk cannot be degraded with high rate of development and as a consequence, undegraded yolk comes to lie in abnormal positions and thereby disturbs normal morphogenesis. On the other hand, when these species were kept at 10°C (just above the lower lethal temperature) the yolk reserves become depleted prior to hatching and embryos died. On the cuticular level, extreme temperatures may induce drastic physiological changes with the result that development becomes abnormal. Silkworm eggs can be induced to undergo thermal androgenesis when exposed to 40°C during the meiotic division. The heat-shock phenomenon has attracted much attention as a model system for the induction of specific genes by exogenous treatment. When *Drosophila* cells are exposed to 37°C, a small number of so called heat-shock genes become activated almost immediately. Cold treatment of embryos may also cause embryonic death or abnormal development. Cold treatment of *Bombyx mori* silkworm eggs leads to formation of tetrapoid individuals which lay large eggs (Kawamura and Nakada, 1981).

Temperature and Food Consumption and Utilization

Variations in the fluctuations of temperature prevent insects from attaining their physiological potential performance and they achieve it only if placed in an ideal and favorable environment. As a consequence of natural selection imposed by less ideal environmental conditions, insects have evolved certain abilities to evaluate their environment and to make decisions involving physiological, behavioral and genetic responses. These responses frequently involve changes in the consumption and utilization of food, rate and time of feeding behaviour, metabolism, enzyme synthesis, nutrient storages, flight behavior and other physiological and behavioral process. Natural environments exhibit large amount of variation in abiotic components (temperature, humidity etc.) which play an important role on the consumption and utilization of food. Variation in environmental factors away from the conditions that allows insects to achieve their ideal performance may reduce their performance unless compensated for by changes in their physiology and behavior. Variable temperature regimes may influence performance differently compared to constant temperature; growth performance is often stimulated in fluctuating temperature regime (Scriber and Slansky, 1981). Insects have also evolved various enzymatic and metabolic adaptations that allow them to survive and develop in a broad range of temperatures. Temperature acclimation, physiological and behavioral thermoregulation allows individual insects to compensate to various degrees for changes in ambient temperature (Heinrich, 1981).

Temperature and Biochemistry of Digestion

In a series of investigations carried out at different constant temperatures on *Tenebric molitor* and *Morimus funereus* beetles, established that long-term adaptation is mediated via the retrocerebral complex (affecting cytological state of cerebral neurosecretory cells) and is a programming of the steady state level of mid-gut proteolytic and amylolytic activities. These two beetle species differ in their temperature requirements *viz., T. molitor* develops well at 23°C and *M. funereus* at 13°C. The effect of temperature along with starvation/ligation, on the digestive physiology of these beetles indicated that retrocerebral trophic factor (stimulating mid-gut protease) is involved in the process, while mid-gut amylase synthesis is apparently independent of this factor. Protease decreases in *Tenebrio* larvae when kept at high temperature and extracts of retrocerebral complex elevate mid-gut protease in ligated or temperature inactivated larvae. These results demonstrate a positive control of mid-gut protease activity, resulting from either reversible inactivation of the cerebral neurosecretory mechanism by temperature or physiological separation from the retrocerebral source as by ligation.

Temperature, Fecundity and Fertility

The rate of egg production varies with temperature; accelerated up to a point and then falls-off rapidly (Mathur and Lal, 1994). But the temperature limits between which reproduction can occur are often much narrower than the range of temperature over and above which the other activities of the same species remain normal. Females of *Locusta migratoria* fail to mature their eggs when the day and night temperature alternate between 30°C and 20°C, *Pediculus* will not lay eggs below 25°C, *Anapheles quadrimaculatus* will not lay below 12°C. The male seems to be more sensitive than females to abnormal temperatures. When *Euchalcidia caryobori* exposed to 16°C for 10 days, the female still laid the total number of eggs, but 70 per cent of males were sterile. In *Drosophila* kept at 32°C, 50 per cent of the females and 96 per cent of males were sterile. These males were able to copulate but no ejaculation of sperm occurred as the sperm in the male organs lost their motility and subsequently degenerated. In the silkworm, *Bombyx mori*, maximum ovulation and fecundity with minimum retention was observed at temperature 25.36 ± 0.17°C (optimum) and any fluctuation from optimum level decreased ovulation, oviposition, fecundity and increased retention of eggs (Mathur *et al.*, 1988).

Temperature and Embryonic Diapause

A critical element of adaptations is how insects respond to the presence and reliability of environmental signals, not just the environmental conditions themselves. Insects use one or a combination of photoperiod, temperature, thermo-period, moisture, food and other factors to assess the current or future suitability of habitats for development or reproduction. The best cues are reliable, frequent and easily recognized (Table 8.3).

The determination of diapause is almost maternal in mulberry silkworm *(Bombyx mori)*, *i.e.* temperature and photoperiod are most efficient at the embryonic stage of previous generation (Table 5.3 and 5.4) and only supplemental in the post-embryonic stages. The sensitive embryonic stages begin just after blastokinesis. Incubation of bivoltine eggs at high temperature results in induction of diapause in the next generation and low temperature incubation results in the production of non-diapause eggs (*c.f.*, Yokoyama, 1973). Incubation as low as 15°C causes production of non-diapause eggs in the next generation, whereas, diapause eggs are induced by incubation at 25°C. When eggs are incubated at intermediate temperature of 20°C, the developmental fate remains undetermined in the embryos. High temperature at younger larval stages and low temperature at late larval stages acts to induce diapause eggs (Tazima, 1978). Egg diapause is regulated by photoperiod as well as temperature during embryonic stage of the female and is completely independent of photoperiod during post-embryonic development. Thus, photoperiod becomes effective in regulation to development only when eggs are incubated at an intermediate temperature. In these eggs, long photoperiod causes induction of diapause and short photoperiod non-arrested state of development. Chilling is generally effective in terminating diapause. Chilling is reported to reduce water absorption by diapause eggs of *Chorthippus brunneus*. In *Leptopterna dolobrata* (mired bug), chilling elevates oxygen uptake in diapause eggs, thereby indicating a gradual transition from diapause to post-diapause (Braune, 1976), and chilling also elevates glycerol content in diapause eggs of melon beetle, *Atrachya menetriesi*.

Table 8.3: Features of Environmental Signals Used by Insects

Element	*Description*	*Sample factors and properties*
Reliable	Correlated with seasonal position and possibility of seasonal change	Photoperiod, thermo-period reliable, food often reliable, other factors mostly unreliable.
Frequent	Available for regular monitoring	Temperature, moisture, photoperiod, thermo-period frequent, other factors mostly intermittent or variable.
Recognizable	Sensors available and seasonal rate of change high	Sensors available for most factors but most factors with low rates of change.

II. Humidity

Water forms a large proportion of insect tissues and survival depends on the ability to maintain and to balance water in the body. There is no limiting range of humidity and most insect can develop at any humidity provided they are able to control their water balance. The water content in insects ranges from less than 50 per cent to more than 90 per cent of the total body weight and there may be much variation within the same species even when reared at identical conditions (Mathur and Lal, 1994). In *Calandra granaria*, the water content is only 46–47 per cent, while in *Talea polyphemus* caterpillar, it amounts to 90–92 per cent . The water content in mulberry silkworm larvae ranges from 77–79 per cent by weight, but adults have only 64–69 per cent water. In *Popillia japonica*, the larvae contain 78–81 per cent of water, the pupae 74 per cent and the adult 67 per cent .

Humidity and Rearing

Abiotic factor that has significant impact on the performance of insects in terrestrial environments is humidity. Humidity interacts with the availability of free water and with the water content of the food. Demands in humidity vary depending on the biological circle. Humidity mostly shows indirect effect on growth and development. In silkworm, humidity influences physiology through withering of leaves and sanitation of rearing beds. Under too dry conditions, the leaves wither very fast and become unsuitable for feed, resulting in retarded growth of larvae which makes them weak and easily susceptible to diseases and other adverse conditions.

Ordinarily high relative humidity during rearing of young age silkworm larvae results in lesser loss of silkworm larvae than low relative humidity. During late age silkworm rearing, low humidity reduces their loss. However, during moult and throughout the moulting phase, the humidity should invariably be maintained slightly lower than in the rearing period. Humidity during rearing phase of 1st, 2^{nd,} 3rd 4th and 5th instar should be maintained at 85 per cent, 85 per cent, 80 per cent, 75 per cent and 70 per cent respectively for healthy growth of larvae and good quality cocoon. During mounting the room should be well ventilated and optimum humidity is to be maintained at 70-75 per cent . During egg incubation, it is important that humidity should be maintained at 80 per cent on an average for normal growth of embryo. If humidity falls below 70 per cent during incubation, the hatching invariably will be low. For regulating humidity in the rearing bed, paraffin paper and wet foam pads or paper bands are used. If humidity is too low, sprinkling of water on the floor may be found useful to increase it.

Humidity and Embryonic Diapause

Ando (1974) classified diapause into three groups based on the requirement of water for embryonic development *viz.*, sufficient water is stored in the egg at the time of oviposition, so that embryogenesis is completed till hatching as in *Bombyx mori*; water is absorbed into egg before diapause establishment and utilized for post-diapause development as in crickets; and water is absorbed mainly after diapause termination for the completion of embryogenesis as in grasshoppers. The mechanism of water absorption takes place through egg shells. In silkworm, *Bombyx mori*, hydrocarbons comprise the major lipid of the egg shell and are believed to take part in water evaporation. Humidity may affect metabolism and hence the rate of development of insects. *Ptinus* eggs take 15 days to develop at 20°C temperature at a relative humidity of 30 per cent, but at 90 per cent humidity, the incubation period is reduced to 10 days. In *Locusta*, larval development is faster at 70 per cent relative humidity, being slower at lower and higher humidity. Humidity affects diapause nature of eggs and voltinism in silkworm only if the incubation temperature is between 15°C–25°C (Table 5.4).

Humidity and Respiration

The effect of humidity on respiration is complicated by a number of factors, most important of which is the water content of the insect. Humidity effects are very closely associated with temperature effects *i.e.* water loss by desiccation, spiracular diffusion, retention of ingested water and production of metabolic water. In alternating wet and dry conditions, *Drosoplila* have a higher respiration in saturated air than in dry air (Mathur and Lal, 1994). Mealworm subjected to various humidifies at temperature from 8°C to 37°C show no variation in CO_2 output. It has been reported that water content of the insects may not be a limiting factor as starved *Papilla* larvae at different relative humidity does not show any effect. Utilization of tissue water is rapid in those insects where lowered respiration and death follows starvation. From these, it can be inferred that changes in humidity have no direct effect on respiration, but indirectly may modify metabolism through temperature, nutrition and metabolic water changes.

Humidity and Survival

The survival at different relative humidity depends largely on the ability of insects to maintain its water content. If it falls too low, the insect dies with few exceptions. If in the eggs and pupa, the insect is unable to replenish its water, the duration of survival will be inversely proportional to the rate of water loss *i.e.*, saturation deficit. The more permeable cuticle shows lower saturation deficit at which the insects will die (Mathur and Lal, 1994).

III. Ventilation

Silkworms like any other animal require fresh air for their various physiological activities. The freshness of air can be determined by its CO_2 contents. Due to respiration, carbonic acid gas is released in the rearing bed and rearing room. Although atmospheric CO_2 content is generally 0.03–0.04 per cent in the rearing room but it can increase due to firewood, coal, smoke of fuel, carbon dioxide produced due to fermentation of leaves on rearing beds, respiration of mulberry leaves, carbon mono-oxide, ammonia and sulphur dioxide etc. and these gases are injurious to growth, development and health of silkworm larvae. When built up of these poisonous gases reaches beyond the tolerance limit of silkworm larvae, they start to show the symptoms of sluggishness and even stop to feed. Young silkworm larvae are more susceptible to the poisonous gases and hence artificial circulation of air is extremely useful in bringing down the temperature and humidity besides removing the poisonous gases from the rearing room. Carbon dioxide content exceeding 1 per cent in rearing room is reported injurious for silkworm larval health and the relation between concentration of CO_2 in a rearing room and mortality of silkworm is linear (Singh and Saratchandra, 2004). Rajan and Himantharaj (2005) stated that air current of 1.0m/sec during 5th age rearing reduce the larval mortality and improves ingestion, digestibility, larval weight, cocoon weight and pupation rate compared to those recorded under no ventilation condition. However, 1 per cent CO_2, 1 per cent formaldehyde gas, 0.02 per cent SO_2 and 0.1 per cent NH_3 in the air in rearing house is safer limit.

IV. Photoperiod

Silkworm is an insect of small day with positive phototactism. It constantly influences the physiology of silkworm. Silkworms are photosensitive and generally have a tendency to crawl towards the dim light. Larvae of silkworm do not prefer either strong light or complete darkness but usually light phase in contrast to the dark phase activates the larvae which prefer light within the range of 15–

30 lux. Patil and Gowda (1986) reported that if silkworm larvae are fed in complete darkness their duration is longer and cocoon quality is poor. Rearing in either complete darkness or in bright light leads to irregularity in growth and moulting. Light phase usually makes larval duration longer than the dark phase. Photo phase for rearing of silkworms should be 16 hrs light per day followed by a scot phase of 8 hrs. It is advisable to rear silkworms in dim light during the daytime and in the dark during at night for healthy growth of larvae. Rearing of silkworms in continuous light delayed growth considerably leading to appearance of pentamoulters and reduced larval and cocoon weights besides cocoon quality.

V. Genetic Variations

Environmental responses, whereby a given genotype allows a range of plastic reactions to different environments, act in concert with genetic variations (Table 8.4). All organisms show individual variations. Traits that are crucial to survival may be canalized so that the important adaptations will not be lost through variations away from the means (Stearns and Kaweeki, 1994). On the other hand, extensive variations make it possible to produce individuals adapted to variable habitats. Such variation tends to be genetic rather than environmental in two circumstances–first, when habitats or at lease different regional habitats are stable and second, when habitats change and the changes cannot be predicted from environmental cues, preventing adequate plastic adjustments.

VI. Food Limitations

Survival when food is limited can be enhanced by eating what is available rather than restricted diet, and by tolerating starvation (Table 8.5). Many predators change diets seasonally, eating any common species of the right size that they can catch. Some herbivorous species also change food plants in seasonal patterns. Many species prolong their development if food is limited or of poor quality and others delay reproduction in the absence of suitable adult food. For example, typical anautogenous biting flies develops eggs only after they acquire a blood meal.

Starvation can be withstood by reducing metabolism. Of course, species in diapause do this but also predators such as spiders almost halve their metabolic rates without dormancy when starved, and may switch to metabolizing fat (Tanaka and Ito, 1982). Such abilities

Table 8.4: Genetic Systems in Insects that Respond to the Nature of Change in Climates or Habitats

Element	*Alternative Features*	*Descriptions*	*Typical Climate or Habitat*
Patterns of genetic variation	Limited	Little variation, normally distributed, canalization	Many
	Extensive	Wide and retained variation	Variable
	Polymorphism	Different morphs in one cohort	Unpredictable habitats without reliable environmental signals
Partheno-genesis	Obligate	Reproduction always parthenogenetic	Severe and variable over the long term
	Dominant	Occasional sexual reproduction	Severe
	Cyclic	Parthenogenesis and sexual reproduction alternate seasonally	Alternating habitats
	Intermittent	Occasional parthenogenesis but usually reproduce sexually	Unknown

Source: Danks, 2006.

Table 8.5: Common Mechanism of Direct Resistance to Food Limitations in Insects

Mechanism	*Sample Systems or Substances*
Use available supplies	
Widen food range	Wide range of food plants.
Widen habitat	Generalized saprophagy in soil habitats.
Eat what is present	Change diet seasonally.
Tolerate starvation	
Reduce energy use	Prolonged development, delayed reproduction, and other resource allocations.
Reduce energy requirements	Reduced metabolic rate.
Store reserves	Accumulate extensive reserves, especially lipids in fat body.

Table 8.6: The Main Features of Trade-offs

Feature	*Components*	*Details*	*Examples*
Limits	Constraints	Resources that are only at maximum levels cannot be traded-off.	Limited food determines final size directly.
	Energy	Energy is the common limiting currency.	Both long-distance dispersal and immediate high fecundity are not usually possible.
	Canalization	Development may be canalized to be inflexible.	Growth rate is insensitive to temperature.
Complexity	Major elements	Certain elements are especially important in most trade-offs.	Customarily emphasized is size or weight, developmental time or rate, growth rate, survival, fecundity, longevity.
	Other elements	Many other resources potentially are traded-off.	Other elements include competitive ability, wing morphs or dispersal, resistance to heat, cold and other adverse conditions, reproductive patterns, egg size verses egg number, diapause duration verses post-diapause fitness.
	Simultaneous trade-offs	Many elements are integrated at once	All of the components listed above are in dynamic balance.
Flexibility	Many pathways	The factors in a potential trade-off can relate in many different ways.	Trade-offs is affected through various metabolic and other pathways.
	Traits below maximum	Developmental and reproductive traits are not necessarily at their maximum levels, even under temporal constraints	Growth rate is not routinely maximized.

are also well developed in species such as ticks and bugs that feed on vertebrate blood. The high arctic moth, *Gynaephora groenlandica* (Lymantridae) also adjusts metabolism according to food supply. Such species can survive for many months without food. A major route for withstanding starvation is the accumulation of extensive reserves, especially lipids stored in the fat body, to supply energy in the absence of feeding.

VII. Trade-offs

Resources are normally limited, but organisms can allocate resources among different traits to maximize fitness. The differential allocation of resources is known as trade-offs which gives insight into how life cycles are adjusted and what types of life cycles are possible in the particular circumstances (Danks, 2006). Resources that are never surplus cannot be traded-off. Allocations essential to survival are protected, so that they are unlikely to be traded-off. Development that is not modified in response to circumstances is said to be canalizes. Trade offs can be made effectively only when there is some stability of the resources being acquired, and they tend to be obscured when conditions are very variable. Conversely, when food resources are especially abundant, the trade-offs normally required may not be necessary. When time to complete development is limited–trade-offs take different forms according to species *viz.*, reduce development time and reduce size; maintain development time and reduce size or increase development time and maintain size (Danks, 2006 a). The main features of trade-offs is summarized in Table 8.6.

Insects survive in seasonal climates by adaptations to resist adversity, including inactivity when appropriate, as well as adaptations to ensure that activity or reproduction coincide with suitable conditions.

Chapter 9

EGG PROTOZOAN DISEASE AND PESTS

(A) Protozoan Disease

The mulberry silkworm, *Bombyx mori,* is prone to infections of various pathogenic organisms. Microsporidiosis of the silkworm caused by highly virulent parasitic microsporidian, *Nosema bombycis* (Nageli) is one of the most serious maladies, which determines the success or failure of sericulture industry in any country. Infections of the disease range from chronic to highly virulent and can result in heavy loss to sericulture industry. Several strains and species of microsporidians have since been isolated from the infected silkworms; the disease is becoming increasingly more and more complex. Epizootiology, development of immunodiagnostic kit, fluorescent antibody technique and use of ideal disinfectant, chemotherapy and thermo-therapy techniques and management strategies has been addressed for identification, destruction, prevention and control of disease causing micro-organisms. Technique of forced eclosion test and delayed mother moth examination also plays an important role in the detection of the disease and to harvest stable cocoon crop. In this Chapter efforts have been made to present annotated information on the causative organism, pathogenesis, manifestation, diagnosis and management of microsporidia infecting mulberry silkworm, *Bombyx mori.*

Pebrine disease of silkworm is a disastrous disease which ensures success, or failure of sericulture industry in any country. This is a chronic insidious disease, evidenced by the historical fact that the rise and fall of the pebrine disease corresponds with ups and downs of the sericulture industry in silk producing countries of the world. At one time or the other, it has severely damaged sericulture in all the silk producing countries of the world. The evidence showed that the pebrine outbreak started in France during 1845. France total production, which had reached 26,000,000 kg in 1853, fell to only 4,000,000 kg in 1865. During this time the disease was so rampant that sericulture in France was on the verge of collapse (Tatsuke, 1971). The history of research on pebrine disease progressed with the advancement of microbiology in 19th century. The name 'pebrine' was coined by De Quadrefages (1860) because of the appearance of pepper-like spots in the diseased larvae. The disease-causing microorganism was first observed in haemolymph of silkworms and was given the name 'Hematozoid' (Guerin–Menaville, 1849). Later Nageli of Germany stated that the disease is caused by a protozoan parasite and named this pathogen, *Nosema bombycis.* Louis Pasteur called the disease as 'Corpuscles disease', made a detailed study on its growth and transmission, discovered that disease is contracted through transovam transmission within the body of mother moth, and suggested the method of preventing the disease. Balblain made it known that the pathogen of pebrine disease belongs to Protozoa. Later on in 1909, Stempell published the detailed of his study on the life history of the pathogen of pebrine disease. Based on the microsporidians unusual cytological and molecular characteristics such as primitive type of nuclear division, devoid of mitochondria, prokaryotic sized ribosome's and ribosomal RNAs, they have been phylogenetically considered as one of the earliest known eukaryotes (Vossbrinck *et al.*, 1987).

In India, the first record of spread of incidence of disease was at the end of the 19th Century in Kashmir Valley. In 1890-1900, the disease swept through Mysore and Madras province. Thereafter, the disease reappeared during 1925-1930 in an epizootic form (Chitra *et al.*, 1975). Disease epidemics were again observed during 1991-1992 which resulted in considerable crop loss and revenue of over 200 crores during the period. Since then the incidence of the disease is being observed intermittently in silkworm crops in different parts of India. The microsporidia are spore forming, small, obligate, intracellular living eukaryote infecting both beneficial and non-beneficial insects (Nataraju *et al.*, 2005). More than 140 genera and 1200 species of microsporidians are recorded from insects and fishes (Canning, 1993, Samson *et al.*, 1999a). Among them at least 200 belongs to the genus *Nosema* (Sprague, 1982) and most *Nosema* species are parasitic to invertebrates. A majority of these such as *N. bombycis* and *N. tyriae* (Canning *et al.*, 1999), *N. mesnili* (Cheung and Wang, 1995), *N. algerae* (Muller *et al.*, 2000), *N. aphis* and *N. trichoplusiae* (Malone *et al.*, 1994) are pathogenic to various insects. The microsporidian infection remains a major threat to sericulture industry with its recurrent occurrence. More than twenty wild insect species are found to have microsporidian spores that can cross-infect silkworm. Pebrine, the spores of microsporidian (*Nosema bombycis)* is one of the most dreaded diseases of the silkworm, *Bombyx mori*. Pebrine infects almost all ages, stages, breeds and hybrids of the silkworm by both transovarial and peroral infection. It is highly infectious and difficult to eradicate after the occurrence of infection. The earliest research on pebrine was confined especially with the epizootiology and prevention of the disease (Weiser, 1969; Ishihara, 1963; Fujiwara, 1979). Microscopical method of mother moth examination, although widely practiced mainly due to its simplicity, does not assure full-proof detection of the microsporidian.

To circumvent this problem efforts have been made to evolve simple, precise and more accurate method of detection of the disease (Geethabai *et al.*, 1985; Fujiwara, 1993; Baig *et al.*, 1992; Shi and Jin, 1997), identification of alternate host (Fujiwara, 1993; Samson, 2000), use of chemotherapy and thermo-therapy for the prevention and control of disease (Hayasaka, 1990) besides identification of intermediary stages (Santha *et al.*, 2001) but with little success. Even though, the researches and fight against the pebrine has been continuing for more than a century, loss due to disease has not been eliminated completely. However, historical evidences suggest a significant relationship between success of sericulture industry and the control of the disease. Therefore, for the improvement of sericulture industry and to save it from the crop losses caused due to this chronic nature of the disease, it is essential to have a full-proof diagnostic and preventive technique.

Systematic Position

Phylum	:	Protozoa
Sub-phylum	:	Cnidospora
Class	:	Sporozoa
Sub-class	:	Neosporidia
Order	:	Microsporidia
Sub-order	:	Monocnidina
Family	:	Nosematidae
Genus	:	*Nosema*
Species	:	*bombycis* (N.)

Pathogen

Several microsporidians are known to cause the disease in silkworm. Among them the most common are *Nosema bombycis* and strains of *Nosema* sp., NIK-2r, NIK-3h, NIS-001, NIS-M11, NIS-M14), *Vermimorpha* (NIS-M12 and NIK-4m), *Pleistophora* sp. (NIS-M27), *Thelohania* (NIS-M32) and *Leptomonas* sp.

Life Cycle of *Nosema bombycis*

The life cycle of *Nosema bombycis* (Nageli) includes three stages namely, spore, planont and meront. Mature spore is oval or ovocylindrical and measures 3–4 microns by 1.5 -2.0 microns with three-layered membrane, the inner, middle and outer. They can be observed at 600 magnifications under a microscope. The spores consists of:

- ☆ Spore membrane, which encloses the sporoplasm.
- ☆ Sporoplasm in the form of a girdle across the width of the spore.
- ☆ Anterior and posterior vacuoles.
- ☆ Two nuclei in the sporoplasm.
- ☆ Posterior capsule.

The spores are highly refractive, appearing light green under the microscope. The outline is smooth and the spores are heavier than the water. The spore belongs to the dormant stage of pathogen and possesses great resistance. For example, they can remain infective after three years in the dried body of the female moth and remain active after being sub-merged in water for five month. The spore germinates in digestive juice of silkworm larva and produces a long polar filament having a length of more than 30 times that of the lengthwise dimension of the spore, on the end of which grows a sporoplasm (germs). The sporoplasm is having two nuclei and other cell organs and possesses the limiting membrane. The sporoplasm multiplies through fission, comes out of the haemolymph through intracellular spaces spreading to every part of the body, and lives in various systems particularly in the fat body and muscular tissue, becomes a nucleus and form spore after multiplication through fission. Spore formation is apansporoblastic, disporous and dimorphic. One type of sporoblast of the long polar tube type turns into single spore and many coil of polar tube. The other type is the sporoblast of the short polar tube, which turns into single spore with few coils of polar tube. Spores of short polar tube hatches directly in the host cell. The mature spore is unicellular endomembranous differentiation of its sporoblast. The pathogen of pebrine disease is capable to complete its life cycle within 4 days. The pathogen parasitizes the ovary first and when 4th or 5th instar larvae after pupation and emergence becomes moth they move into egg and after deposition of eggs undergo multiplication and develops into a disease in the embryo or in the body of silkworm in next generation. After the deposition of eggs the pathogen grows and multiplies within the egg. The mature spore is unicellular endo-membranous differentiation of its sporoblast (Vavra and Maddox, 1976). These authors designated the sporoblasts as Phase-I sporoblasts and Phase-II sporoblasts. The Phase-I sporoblasts are characterized by the presence of a dark staining spherical body.

Source and Stage of Contamination

Transversally infected seeds are the primary source of contamination. Contaminated seed crop rearing and grainage buildings, appliances, silkworm litter, mulberry leaf fed to the silkworm harbored by infected insects etc also contributes to the spread of the disease. The incidence of pebrine varies with the breeds/hybrids of silkworm, the developmental stage and the rearing environment. Resistance to pebrine is greater in Chinese breeds, less in Japanese breeds and least in European breeds (Govindan *et al.*, 1998) and that multivoltine breeds are relatively more resistant than bivoltines (Patil and Geethabai, 1989). Young silkworms, newly moulted and starving larvae are susceptible and show high mortality. In India, Nistari and C. Nichi silkworm breeds are more resistant compared to others. Patil and Geethabai (1989) reported that among bivoltines breeds NB7 is most susceptible followed by NB4D2, KA and NB18. Although the disease resistance appears to depend on the genetic constituents of a particular breed, nevertheless factors such as pathogen load, inadequate nutrition and the environment in which insects are reared may also affect resistance (Patil, 1993). In addition, the physical and physiological characteristics of hosts may make the invasion of microsporidians possible (Weiser, 1969, 1977). The larvae infected during 1st and 2nd instars show normal growth up to 3rd instars. Disease symptoms appear during later half of 4th instars to first half of 5th instars and die before spinning. If the contamination takes place in 3rd instars, the larvae show symptoms of the disease in the late 5th instars and die on the mountage before cocooning. These larvae discharge spores through faecal matter during 4th and 5th instars. If these larvae are reared with healthy larvae, the spore discharge by infected larvae provides the source of contamination and digestion of spores by healthy silkworms results in spread of the disease. This stage of contamination is known as 'second stage of contamination'. Larvae infected during 4th and 5th instars pupate and on emergence lay contaminated eggs. This phenomenon is referred as 'transovarian transmission'. Most of the larvae infected through transovarian transmission show irregular moulting and growth, becomes tiny or under grown and die before 3rd moult after discharging spores. The contamination occurring from transovarially-infected larvae is termed as 'first stage of contamination'. The minimum number of spores required for contamination through per oral infection varies with each instar. Iwano and Ishihara (1981) stated that 1–10 spores are sufficient to cause disease in 2nd instars larvae, while approximately 100 such spores are required in 5th instars for the same symptoms to occur. Transovarian transmission is 100 per cent in case of *N. bombycis* and only 1.2 per cent with *Nosema* sp. M11 (Han and Watanabe, 1988).

The spores of different microsporidia infecting silkworms differ in their morphological characters, some are larger than mature spore and some are long, thin and pear-shaped with different size, shape and luster. Sometimes the conidia of green muscardine and red muscardine bear a striking resemblance to the spore of pebrine disease. Horizontal transmission of pebrine spore is possible through contaminated rearing bed, contaminated mulberry leaf and through contaminated layings (Govindan *et al.*, 1998). Baig *et al.* (1988a) reported that spread of disease in rearing trays is also dependent on the density of diseased silkworms. Growth and multiplication of pathogen are influenced by growth of its host. When egg enters into diapause, the growth and multiplication of pathogen stops simultaneously and when egg starts growing, the pathogen also starts growing and multiplying.

Cross Infectivity

Different species of insects known to carry microsporidians causing cross-infectivity to silkworms are found harboring in and around mulberry garden. Enormous quantity of microsporidian spores is observed in *Catopsilia* sp., an inhabitant of mulberry garden (Kishore *et al.*, 1994) and found infective to silkworms. Butterflies causing microsporidian infections to silkworms are also reported (Samson *et al.*, 1999a, b). Singh *et al.* (2007) reported that butterflies *i.e. Eurena hecabae* and *Zizina otis* carry microsporidian spores infective to silkworms. These insects are potential source of contamination as spores of pathogen is excreted along with litter on mulberry leaves in the garden and when these leaves are fed to silkworms, they causes the disease to appear.

Physiological Stability

Generally, large number of factors *viz.*, temperature, humidity and abiotic components of the substrate influence the survival of microsporidians (Kramer, 1976). The spores belong to the dormant stage of pathogen and possess great resistance, can remain infective after 3 years in the dried body of the female moth, and become active after being submerged in water (Li, 1985). When kept in dark, the spores are reported to remain viable for as long as seven years, but when spores are directly exposed to sunshine (Anonymous, 1980) they remain viable for 6–7 hrs and when treated with hot water survive for just 5 minutes. Studies conducted on the viability of pebrine spores in soil and compost under tropical conditions showed survival of spores for maximum period of 225 days in wet soil and minimum of 135 days in wet compost (Patil, 1993). Srikanta (1986) observed that spores remained infective even after 150 days of refrigeration and after 90 days in moist soil and faeces. He further stated that the viability of spores is lost in 60 days in dry soil and in 5 days when stored at room temperature. Resistance of spores to different disinfectants indicates that it can remain viable for 10–30 minutes in the solution of corrosive sublimate, for about 5 hrs in formalin and 10 hrs in chlorinated lime solution (diluted 10,000 times). Bleaching powder containing 1 per cent and 3 per cent active chlorine can render spore inactive in 30 minutes and 10 minutes respectively. When the degree of infection is relatively high, the egg often becomes sterile or dead, but when the contamination is of low degree, the egg hatches and the disease develops at larval stage and caused death of larvae at later stages of development. The growth and multiplication of pathogen in the eggs are influenced by the growth of the host. When the egg diapauses, the growth and multiplication of the pathogen stops simultaneously and when the eggs start growing by incubation, the pathogen also starts growing and multiplying.

Alternate Hosts

Most microsporidians prefer to have alternate hosts because of many advantages for them *viz.*, dispersal, transmission and survival. The perpetual incidence of microsporidian infection in silkworms may be due to various sources of secondary contaminations including alternate hosts in and around mulberry garden. In addition to *N. bombycis*, seven other microsporidians belonging to the genera *Nosema*, *Pleistophora*, *Thelohania*, *Vairomorpha* and *Leptomonas* spp. has been isolated from the silk moth (Govindan *et al.*, 1998). They differ in their spore morphology, target tissue and virulence and have been designated as M11, M12 and M14 (*Nosema* sp.), M24, M25, M27 (*Pleistophora* sp.) (Fuziwara, 1984a & b) and M32 (*Thelohania* sp.) (Fujiwara, 1985) (Table 9.1). Three microsporidians designated as NIK-2r, NIK-3h and NIK-4M have been isolated from Karnataka (India) and these are immunologically dissimilar to *N. bombycis* (Ananthalakshmi *et al.*, 1994).

N. bombycis is also reported to infect *Samia Cynthia ricini* and Indian tropical tasar, muga and Chinese tasar silkworms (Talukdar, 1980). *N. bombycis* is also found to infect a several other lepidopterans like *Spodoptera exigua*, *S. litura*, *Diaphania pulvurentalis*, *Pieris rapae*, *P. brassicae* etc. Veber (1958) reported 32 species of lepidopterans known to develop infection to the peroral inoculation of *N. bombycis* spores. They include *Chilo suppressalis*, *Pieris rapae*, *P. brassicae*,

Spodoptera exigua, S. litura, S. maurilia, Balataea funeralis, Cruptophlebia illepida, Exartema mori, E. morivirum, Diaphania pyloalis, Mycalesis gotoma, Abracus miranda, Descorba simplex, Boarmia selenlia, Menophra atrilinecta, Elydna nonagrica, Otosema odera, Perigea illecta, Plusia chalcites, Pseudaletia unipuncta, Stilpnotia lubricipeda, S. imparilis, Callimorpha quadripunctata, Thaumetopoea processionea, Malacosoma neustria, Gastropacha quercifolia, Lasiocampa quercus, Bombyx mandarina, Antheraea pernyi, A. yamamai, Sphinx ligustris, Agrotis ipsilon, Agrius cinagulatus, Pholera assimilis, Acronicta major, Acrotomycis aceris and *Achaea janata* (Samson *et al.*, 1999b; Kawarabata, 2003; Singh *et al.*, 2007). The lawn grass cutworm, *Spodoptera depravata* serves as a natural reservoir for the pathogen (Ishihara and Iwano, 1991), which shares the surface specific antigens with *N. bombycis and* results in transovarial transmission with less virulence.

Table 9.1: Different Types of *Nosema* Spores

Microsporidian	*Spore Size (µm)*	*Site of Infection*	*Virulence*
Nosema bombycis	3.8 x 2.2	Systemic	High
Nosema sp. (M11)	3.9 x 1.9	Various tissues	Low
Nosema sp. (M12)	4.2 x 2.7	Various tissues	Low
Nosema sp. (M14)	5.1 x 2.0	Various tissues	High
Pleistophora sp. (M24)	2.7 x 1.6	Mid gut	Low
Pleistophora sp. (M25)	3.2 x 1.8	Mid gut	Low
Pleistophora sp. (M27)	5.4 x 3.0	Various tissues	Low
Thelohania sp. (M32)	3.4 x 1.7	Muscles	Low

Epidemiology

Source of Infection

The source of pebrine infections are many including diseased larvae and wild insects affected by the diseases, excreta and faeces, urine of the ripe larvae and moths, discarded eggshells, pupal skin and scales, epicuticle and cocoon shell.

Routes of Infection

Nosema sp. transmits infection both vertically and horizontally in the silkworm. The primary infection is by vertical transmission with the pathogen transferring the infection directly from the parent to the offspring's or by horizontal transmission where the transmission is by secondary infection from one individual to another which is accompanied by ingestion of spores. These two types can also be termed as oral and transovarian transmission. Orally eating mulberry leaves contaminated by pathogen infects silkworm. Trans-ovarian transmission occurs when the pebrine pathogen infects the 4th and 5th instars larvae and then invades the epithelial cells of the ovaries, from where the parasites are transferred to the oogonia, oocytes and nutritive cells. Parasitism of the oocytes results in the death of the eggs. The result of the embryonic infection, however, differs according to the stage when invasion of the embryos occurs. If infection takes place during the formation of embryos, then no embryonic development takes place and the eggs die.

Transmission within the rearing tray is caused by sick silkworms in which large quantities of pebrine spores formed are excreted with the faeces or attaches to epicuticles, thus contaminating the rearing tray and transmitting the disease to the healthy worms. Infection inside the rearing tray may be divided into two categories *i.e.* primary and secondary contamination. Primary contamination refers to that occurring in the 1st and 2nd instar and the excretion of spores taking place in the 3rd or 4th instars. Ingestion of these spores by other healthy larvae constitutes secondary contamination. Secondary contaminated silkworms are able to feed normally and become adult moths but they lay infected eggs. Young silkworms infected in the 1st or 2nd instars usually die in the 3rd instars but rarely in 4th instars. If infection occurs in 4th instars, development may proceed to the adult stage with laying of eggs but the eggs laid will be infected with pebrine. Young silkworms freshly moulted and starving larvae are more susceptible to the disease and show high mortality. The spread of pebrine infection is generally by secondary contaminations.

Symptoms of the Disease

The disease symptoms depend on the parasitism of the metamorphic stage. The symptoms of the disease can be observed in all the stages of silkworm development *viz.*, egg, larva, pupa and moth.

Symptoms in Egg Stage

- ✰ Poor egg numbers.

- ✰ Pilling of eggs one over the other.
- ✰ Lack of adhesive fluid and poor adherence to egg sheet.
- ✰ Lack of uniformity in egg shape.
- ✰ More unfertilized and dead eggs.
- ✰ Poor and irregular hatching.

Symptoms in Larval Stage

- ✰ Larvae show poor appetite.
- ✰ Retarded growth and development leading to un-uniformity in size (unequal larvae).
- ✰ Larvae show altered colour, size, form and activity.
- ✰ Diseased larvae are comparatively paler and translucent.
- ✰ Larvae moult irregularly and show sluggishness.
- ✰ Body shows wrinkled skin.
- ✰ They lose appetite, retard in growth, sometimes vomit gut juice, shrink in size and become undersized.
- ✰ The affected gut becomes opaque and the silk gland shows white pustules in different places along its length.
- ✰ Black irregular pepper like spots on the skin.
- ✰ Irregular and incomplete moulting.
- ✰ The dead larvae remain rubbery for longer period and then turn black.

Symptoms in Pupal Stage

- ✰ Pupae are flabby and swollen with lusterless and softened abdomen.
- ✰ Irregular black spots near the rudiments of the wing and abdominal area.
- ✰ Highly infected pupae fail to metamorphose into moth.

Symptoms in the Moth Stage

- ✰ Delayed and irregular/erratic moth emergence.
- ✰ Clubbed wings with distorted antennae.
- ✰ The diseases moths will be inactive and do not mate properly.

- Scales from the wings and abdominal area easily falls-off.
- Clumpy egg laying.

Spore Isolation and Purification

Isolation, purification and identification of spores from the host are the first step in the study of pebrine disease and its management. Following centrifugal procedure (Fujiwara's method) of microscopical examination is carried out for spore isolation and identification in order to monitor the disease:

- Take 20 larvae/pupae or moth in a cup of domestic mixer.
- Add 80 ml of 0.6 per cent potassium carbonate solution (K_2CO_3).
- Homogenize the larvae/pupae or moth for 1–2 minutes at 10,000 r.p.m. This ensures liberation of spores from the tissues into the medium facilitating more accurate detection of spores during microscopical examination.
- After homogenizing, the contents (homogenate) are transferred into a glass beaker and are allowed to settle for 2–5 minutes. This facilitates to separate host tissues from the liquid portion of homogenate for easy filtration.
- Filter the homogenate by using absorbent cotton or muslin cloth. Volume of filtrate recovered will be around 70 ml.
- Transfer the filtrate into 100 ml capacity of centrifuge tube through funnel. All the centrifuge tubes should be evenly balanced, capped and loaded into the centrifuge machine.
- Centrifugate the filtrate at 3000 r.p.m. for 3–5 minutes to facilitate sedimentation of spores.
- Remove the centrifuge tube from the centrifuge machine.
- Discard slowly the supernatant solution.
- Dissolve the sediment by shaking with a few drops (2–3 drops) of K_2CO_3 or by using a cyclomixer (10–15 seconds) or with a glass rod, which facilitates uniform dissolving.
- Take this smear from the centrifuge tube on the micro-slide, put the coverslip and examine under 600x magnification of microscope.

- ✩ Minimum 5 microscopic fields should be examined per smear to detect the presence of pebrine spores.
- ✩ Two moth testers to ensure the detection of pebrine spores may perform microscopical examination.
- ✩ If pebrine spores are detected, the entire batch of eggs laid by the infected mother moths is to be destroyed by burning.
- ✩ The intensity of spores can be graded as below (Table 9.2)

Table 9.2: Intensity of Spores and Pebrine Infection Grading

Number of Spores/Field	*Grade*
1–3	±
4–10	1 +
11–30	2 +
31–100	3 +
101–300	4 +
301–above	α

However, serological and biochemical studies of microsporidians require high degree of purity. Gochnaner and Margetts (1980) described rapid method for concentrating *Nosema* spores based on continuous flow centrifugation method. Another method based on 'Brownian movement' was also reported. Sato and Watanabe (1980) purified spores using sucrose and percol gradient centrifugation method and reported that centrifugation using percol at 73,000g for 30 minutes resulted in 3 bands:

1. A sharp band consisting of tissues of silkworms, mulberry leaves and bacteria etc.
2. A dim band consisting of mature but inactive spores,
3. Another sharp band consisting of only mature and active spores

Approaches for Pebrine Prevention and Management

Pebrine has remained a threat to sericulture industry since time immemorial. The disease has become more complex now because of the occurrence of different types of microsporidians infecting the silkworm. Some of them belong to other genera like *Vairomorpha* and *Thelohania* and exhibit differences in their pattern of infection

(Samson, 2000). Apparently, the biology of the pathogen has been used as a basis in disease control. The disease is transmitted horizontally by ingestion of spore and vertically by transovarian transmission. This unique characteristic of the disease made difficult to completely eliminate it from the silkworm crops. The earliest method suggested by Pasteur based on selection of pathogen free eggs through careful systematic examination of mother moths for pathogens, after egg laying has been one of the most effective methods even today to avoid the disease in the silkworm crops.

Proper monitoring and testing of seed crops at every successive stages of progress of the crop is to be ensured to produce pebrine free seed cocoons for commercial seed production. Destruction of infected crops as soon as noticed infection is another important step towards pebrine disease management. Since the disease is seed borne, the surface sterilization of eggs immediately after egg laying and also before incubation should be followed to prevent disease occurrence from surface contamination (Singh *et al.*, 1992). Several reports documented the efficiency of thermal treatment of silkworm eggs in minimizing pebrine infection (Bedniakova and Vereiskava, 1958; Fujiwara and Kagawa, 1984; Hayasaka, 1990). The maximum lowering of infection rate is reported in eggs incubated during first two days of their development to 44°C. Singh and Saratchandra (2003) stated that incubation of eggs at higher temperature within 3 days of laying results in significant reduction in pebrine disease. Thermal treatment combined with hydrochlorization to achieve dual objectives of elimination of pebrine and termination of diapause is also reported. Liu *et al.* (1971) achieved remarkable success in reducing pebrine infection after treatment of silkworm eggs at 47°C for 10–20 minutes. Chowdhary (1967) suggested exposure of cocoons to high temperature (33.8°C) at the time of pupation for 16 hrs a day, at 55–65 per cent humidity tends to reduce infection in the resulting eggs. Sheeba *et al.* (1999) reported that thermo-therapy of 7 days old pebrinized cocoons at 36°C for 16 hrs tends to reduce pebrine infection significantly without affecting the growth and development of larvae.

Certain insect hosts tolerate high temperature than their microsporidian parasites and the host can be freed of the disease by rearing the infected individuals at higher temperature until the disease is cured. Attempts have been made by several workers to control pebrine infection in silkworm eggs by temperature treatment. Ovanesyan and Lobzhanidze (1960) and Austrurov *et al.* (1969)

attempted hot water treatment of pebrinized eggs and reported sharp decrease in the degree of infection. Smyk (1959) expressed varying success with hot water treatment. Fujiwara and Kagawa (1984) reported that the parasites in non-diapausing eggs are more sensitive to hot water (46°C for 4 minutes) treatment and there is no harmful effect of the treatment on the normal development of silkworm embryos. However, these methods are not effective enough to eliminate infection completely. Chemotherapy of microsporidians infection has been attempted in silkworm. Of the several therapeutic drugs, Benomyl, Nosematol, Bavistin and Thiophanate have been identified as anti-microsporidian agents to control *N. bombycis* infections (Chandra and Sahakundu, 1983 and Alenkseenork, 1986). The toxins of *B. thuringiensis* are also reported efficient in killing the Nosema spores. Fumagillin is commonly used to reduce the economic losses due to *Nosema*. Though these methods have proved experimentally effective in reducing the multiplication of spores but further studies has clearly showed that they cannot eliminate transovarian transmission significantly. *N. bombycis* is made to be inactive by hilite (Potassium dichloro isocyanurate) (Iwano and Ishihara, 1981). Baig *et al.* (1988b) studied comparative efficacy of four disinfectants *viz.*, hilite, sodium hypochlorite, bleaching powder and formalin in four concentrations (0.5 per cent, 1 per cent, 1.5 per cent and 2 per cent) as surface sterilents against the spread of pebrine disease in a colony of silkworm hatched from surface contaminated layings and reported that all the tested concentrations were effective in preventing the spread of the disease and were also effective in inactivating spores of *N. bombycis* when exposed to 5, 10, 20 and 30 minutes respectively. Kagawa (1980) studied the efficacy of formalin as disinfectant against pebrine and reported increased death rate of spores with increase in formalin concentration and temperature. Iwano and Ishihara (1981) tested nine types of chemicals as inhibitory agent against *N. bombycis*, with high degree of inhibitory effect on the spores. Treatment of silkworm eggs with HCl of 1.03–1.09 specific gravity at 47°C for 10–20 minutes is known to reduce the disease incidence by 97.4–100 per cent (Liu and Zhong, 1988). Hot air treatment (48–50°C) of 12–18 hrs old silkworm eggs also inhibited the development of microsporidians. Silkworm eggs of 36–60 hr old treated with hot water at 46°C for 90–150 minutes, 48°C for 50–70 minutes and 52°C for 4 minutes also inhibited the development of pebrine disease.

However, these methods attempted to control pebrine disease by several workers were found to have limited success. Therefore, development of better and more reliable diagnostic methods to detect pebrine during seed production and silkworm rearing has always remained one of the important and valid strategies to eliminate the disease from silkworm crops. Relatively recently recommended delayed mother moth test is a significant step in the area of pebrine disease diagnosis by microscopic test. In this method, the female moths after oviposition are preserved alive at room temperature for a period of 3–4 days before subjecting for microscopic test. This allows improved sporulation of the pathogen facilitating easy and more accurate detection of the disease (Samson, 2000). It has been reported that the rate of multiplication of *N. bombycis* increases substantially with the age of moths and cephalothoracic region had the highest spore conc ntration, especially around the wing and wing muscles (Sasidharan *et al.*, 1994) and therefore, testing of silk moths 3–4 days after oviposition would be more effective method to detect pebrine with better accuracy. An improved testing method has also been recommended for better detection at egg stage. A sample of egg is incubated at a moderately higher temperature of 32 ± 1°C for 48 hrs to enhance sporulation of *N. bombycis*. Testing of such eggs therefore enhances the chances of detection of the disease. On these lines, even diagnostic techniques based on the principles of immunology were also attempted in several countries including India for detection of pathogen and spore identification, but with only limited success (Baig *et al.*, 1992).

N. bombycis and closely related spores were diagnosed with antibody-sentisized latex agglutination technique (Hayasaka and Ayuzawa, 1987), slide agglutination technique (Li, 1985; Baig *et al.*, 1992), the use of ELISA procedures (Kawarabata and Hayasaka, 1987), fluorescent antibody technique (Sato *et al.*, 1981, 1982), serological technique (Grobov and Rodionova, 1985) and SPA co-agglutination technique (Mei and Jin, 1998) etc. Development of monoclonal antibody techniques, which has very high specificity and stability, has played great role in the study of classification and identification of specific microsporidians (Chen *et al.*, 1989; Carlos *et al.*, 1996). Ke *et al.* (1990) raised monoclonal antibodies against *N. bombycis* spores and applied them to identify pebrine and other closely related microsporidian spores infecting silkworms using ELISA procedure. Shi and Jin (1997) reported that agglutination test using N5 McAb (hybridoma cell lines secreting monoclonal antibody) sensitized latex particle is a very practical technique for the diagnosis of pebrine disease. A simple dipstick immunoassay method tried later for diagnosis of pebrine was also unsuccessful in the field. A simple negative staining procedure (Geethabai *et al.*, 1985) and an immunoperoxidase staining procedure (Han and Watanabe, 1987; Kawarabata and Hayasaka, 1987) have been developed for the clarity during examination of spores. Sironmani (1997) has developed a Western blot method to identify the microsporidian infection and observed that immunological reaction with *N. bombycis* infected silkworm larvae and eggs showed the presence of 17-kDa polypeptide, which is specific to infection. He further reported that 17-kDa polypeptide can be used as a virulent marker for the identification of microsporidian infection. DNA based probes have also been developed for identification of *N. bombycis* (Malone and McIvor, 1995).

Several PCR methods based on amplification of rRNA gene fragments are available for the diagnosis and species identification of insect microsporidia (Kawakami *et al.*, 1995, 2001). Molecular techniques developed have more sensitivity and specificity in the detection of the disease (Hatakeyama and Hayasaka, 2001). Nageswararao *et al.*, 2004) have studied the pathogenesis, mode of transmission, tissue specificity of infection and SSU-rRNA gene sequences for the microsporidian isolates from the silkworm, *Bombyx mori*. Genetic characterization and relationship between different microsporidia infecting the mulberry silkworm, using inter simple sequence repeat PCR (ISSR-PCR) analysis have been reported (Nageswararao *et al.*, 2005). They differentiated six different microsporidians through molecular DNA using ISSR-PCR and stated that ISSR-PCR analysis in future may emerge as a powerful tool to detect, diagnose and identify microsporidians, which are difficult to study with microscope because of their extremely small size. A new technique based on identification of intermediary stages has also been suggested for diagnosis of pebrine (Santha *et al.*, 2001).

Immunodiagnostic Assay in Silkworm

Various diagnostic immunoassays have been developed and employed for the detection of silkworm different diseases of the silkworm (Table 9.3).

Table 9.3: List of Immunoassays Tested for Identification/Detection of Silkworm Pathogens

Sl.No.	*Immunoassay*	*Pathogen*	*References*
1.	Precipitin test	*Bm*CPV, *Bm*NPV	Miyajima, 1981
		*Bm*DNV	Arakawa and Shimizu, 1985
		*Bm*IFV	Sekijima, 1971
2.	Immunodiffusion test	*Bm*DNV	Arakawa and Shimizu, 1985 and 1987; Chen, 1988; Seki, 1984
		*Bm*IFV	Seki and Sekijima, 1976, Shimazu, 1982
3.	Neutralization	*Bm*CPV	Miyajima, 1981
	test	BmIFV	Kurisu *et al.*, 1975
4.	Haemaggluti-	*Bm*CPV	Miyajima and Kawase, 1969
	nation test	BmIFV	Sato and Kawase, 1971
5.	Latex	*Bm*CPV	Shimizu and Arakawa, 1986
	agglutination test	*Bm*DNV	Arakawa and Shimizu, 1985, 1987; Shimizu *et al.*, 1991
		*Bm*IFV	Shimizu *et al.*, 1983; Sivaprasad *et al.*, 1997
		*Bm*NPV	Arakawa, 1989; Nataraju *et al.*, 1994.
		N. bombycis	Hayasaka and Ayuzawa, 1987; Baig *et al.*, 1992; Sengupta, 1993
6.	Fluorescent	*Bm*DNV	Maeda and Watanabe, 1987
	antibody test	*Bm*IFV	Sato *et al.*, 1978; Shimizu *et al.*, 1983
		*Bm*NPV	Nagamine *et al.*, 1991
		N. bombycis	Sato *et al.*, 1981; Kobayashi and Yamazaki, 1987
7	Enzyme Linked Immuno Sorbent	*Bm*CPV	Mike *et al.*, 1984; Ito *et al.*, 1985
	Assay (ELISA)	*Bm*DNV	Shi and Ding, 1989; Arakawa and Shimizu, 1985
		*Bm*IFV	Shimizu, 1982; Sivaprasad, 2003

Contd...

Table 9.3–Contd...

Sl.No.	Immunoassay	Pathogen	References
		*Bm*NPV	Tau *et al.*, 1990; Nagamine *et al.*, 1991
		N. bombycis	Kawarabata and Hayasaka, 1987; Miyamoto, 1989; Ke *et al.*, 1990
8.	Dipstick Immunoassay	*Bm*NPV	Nataraju *et al.*, 1994.
		*Bm*IFV	Nataraju and Datta, 1999; Sivaprasad *et al.*, 2003

Though these tests are simple and sensitive, unless standard methods are evolved for their effective field applicability, they cannot create any impact on pebrine disease diagnosis in the field. To maintain the quality of silkworm eggs, several attempts have been made to improve the sampling procedure from time to time (Kurisu, 1986; Kurisu *et al.*, 1985; Fujiwara, 1993). Also, procedures have been developed for detection of pebrine spores in soil/dust, rearing and grainage houses, on mulberry leaves, eggshells/unhatched eggs, litter etc. (Singh and Saratchandra, 2004). The sample size for examination of faecal matter to detect presence of pebrine has been described by Patil *et al.* (2001). As it is not possible to examine all emerging moths in the commercial grainages, Fujiwara (1993) suggested 20 per cent sampling method and reported probability of detection of the pebrine disease (Table 9.4).

Destruction of disease-causing microorganisms at various levels is general method of preventing and controlling the disease. Surface sterilization of disease free layings, maintenance of strict sanitation, hygienic rearing, frequent and careful examination of stock, disinfections of rearing rooms and appliances, removal of dead and infected larvae to be adopted strictly to get-rid-off the disease. Exposing all the contaminated materials and equipments to direct sunlight, disinfections with 2 per cent formalin solution or 5 per cent bleaching powder solution is the most effective and simple eradication method of the disease. However, the pathogen killing action of the disinfectants is influenced by several factors such as temperature, humidity, concentration of disinfectants and duration of treatment (Kagawa, 1980). Recently, a new disinfectant chlorine dioxide (Serichlor) has been considered as an ideal disinfectant for all the types of rearing/grainage houses. In combination with slaked lime, it is 2.5 times stronger than chlorine and 2 times stronger than sodium-hypochloride. It is least corrosive and non-hazardous. When no single technique is sufficient to check the disease in field, it becomes obligatory to choose a multi-pronged approach. However, the techniques only help in detecting the disease and the only way out is to destroy the diseased silkworm crops, which causes loss and efforts are to be made at all levels for the prevention of the disease.

Table 9.4: Probability of Detection of Pebrine in 20 per cent Sampling Method (Index)

No. of Egg Cards (20 layings on each card)	*Population (No. of layings)*	*Pebrine*	*Samples*	*Probability*	
				Non-detectable	*Detectable*
20	400	2	80	0.6400	0.3600
30	600	3	120	0.5120	0.4880
40	800	4	160	0.4096	0.5904
50	1000	5	200	0.3277	0.6723
60	1200	6	240	0.2621	0.7379
80	1600	8	320	0.1678	0.8322
100	2000	10	400	0.1074	0.8926
150	3000	15	600	0.0352	0.9648
200	4000	20	800	0.9885	0.9885
250	5000	25	1000	0.9620	0.9620
300	6000	30	1200	0.9988	0.9988
500	10000	50	2000	1.0000	1 .0000

Rate of pebrine infection = 0.5 per cent in female moths.

Source: Fuziwara, 1993.

A burning problem in the field of microsporidiosis is the increasing number of different microsporidians that are being encountered in silkworm crops (Fuziwara, 1980 and 1993). These microsporidians have shown to exhibit varying degree of virulence and many of them, though infective and pathogenic, have demonstrated low multiplication rate in the silkworm. Some of them have not shown vertical transmission in the host. However, as of today, there has been no specific testing procedure to discriminate these microsporidians in the field to take appropriate action while preparing disease free silkworm seed. If pebrine is to be controlled effectively, a system has to be evolved where either a seed cocoon grower or a seed producer is not put in hardship due to reoccurrence of the disease.

Approaches for Control of Pebrine Disease

- Produce healthy eggs to avoid embryonic infection. This can be achieved through scientific processing and inspection method of mother moth examination in silkworm egg production center.
- Dead eggs, dead larvae, dead pupae in the cocoon, dead moths, excrement of larvae from infected trays, exuviae of infected larvae and other possible sources of infection such as contaminated litter should be removed and destroyed.
- Conduct effective disinfections of grainage rooms, equipments, seed production units and the surroundings.
- Maintain hygienic conditions not only during seed crop rearing but also during egg production.
- Laying should be surface sterilized with 2 per cent formalin for 10 minutes before incubations.
- Inspection of eggshells and sample larvae by standard procedure during each instar for possible presence of pebrine.
- Immediate destruction of infected crops, if noticed pebrine in 1st, 2nd, 3rd and 4th stages of rearing.
- If pebrine is noticed in late final stage (5th stage) or in the cocoon stage, the cocoons are to be sent for reeling.
- If pebrine is noticed after purchase of seed cocoons or in moth stage, the eggs prepared should never be supplied and burnt forthwith.
- Synchronization of rearing to facilitate effective monitoring.
- Monitoring and examination of unequal larvae in the rearing.
- Implementation of seed legislation act in true sense.

- ✰ Rearing and grainage operations in same place should be discouraged.
- ✰ Practice of hiring appliances should be discouraged.
- ✰ Integrated approach is the need of the day for proper disease management in sericulture.
- ✰ Destruction of disease causing microsporidians at various levels is the general method of preventing and controlling the disease.
- ✰ Never pass the silkworm seed which has not been subjected to microscopic examination and has come out disease free in true sense.

Forced Eclosion Test

Sample cocoons from a lot are selected randomly and kept at constant temperature of 33°C to facilitate 1 or 2 days early emergence. The emerged moths are examined and if the lot is noticed infected, are rejected immediately.

Delayed Mother Moth Test

Delayed mother moth test is a significant step in the area of pebrine disease detection by microscopic test. In this method the mother moths after oviposition are collected in groups in perforated cardboard boxes/covers and preserved alive. Alternatively, they can be left on dummy sheets in the oviposition trays itself. The boxes/oviposition trays are properly numbered as per egg sheets and preserved in well-ventilated room at ambient room temperature (25–30°C) for a period of 3–4 days before subjecting for microscopic test. This enhanced sporulation of the pathogen in older moths facilitating easy and more accurate detection of the disease. After stipulated period, moth testing is carried out as per recommended procedure in-vogue. By this method, due to enhanced sporulation in older moths, easy and effective detection of pebrine disease is possible. Even under moderately low infection levels, pebrine can be detected by this method. This technique is very useful during basic seed multiplication and production of P1 seeds. It has been also reported that the rate of multiplication of *N. bombycis* increases substantially with the age of moth and cephalothoraxic region had the highest spore concentration, especially around the wing and wing muscles (Sashidharan *et al.*, 1994) (Table 9.5) and therefore, testing of silk moths 3–4 days after oviposition would be more effective method to detect pebrine with better accuracy.

Table 9.5: Sporulation Rate of *Nosema bombycis* in Different Tissues After Emergence of Moths of Silkworm (after Sashidharan *et al.*, 1994)

Body Parts	*Breeds*	*Quantity of Spores on Different Days After Emergence ($x10^7$/gm wt of tissue)*				
		0 hrs	*24 hrs*	*48 hrs*	*72 hrs*	*96 hrs*
Whole moth	PM	4.39	4.50	5.67	21.90	25.50
	NB18	5.92	6.34	12.40	22.00	28.70
Cephalothorax	PM	8.20	10.50	9.40	35.80	44.00
	NB18	7.10	10.20	14.70	38.40	40.10
Abdomen	PM	1.49	2.60	5.50	14.90	21.60
	NB18	5.02	3.80	7.94	14.00	20.30
Wing	PM	6.30	8.50	11.00	25.00	31.30
	NB18	8.61	12.80	24.45	28.60	34.60
Gut	PM	9.62	10.81	10.60	24.60	22.60
	NB18	8.11	8.94	12.40	20.00	21.20
Fat body	PM	0.19	10.15	0.10	0.20	0.20
	NB18	1.34	2.17	2.10	1.77	2.41

Disposal/Disinfection of Materials Used in Testing

Refuse derived from the processes of examination *viz.*, homogenate solution, cotton, muslin cloth etc. should be dumped into a pit after disinfection with 2 per cent bleaching powder solution or dumped and disinfected. Immerse the mixie cups, beakers/ tumblers, centrifuge tubes, glass slides, cover-slips, glass rods etc. in 2 per cent bleaching powder solution for 15 minutes followed by washing with detergent. Cover-slips once used should not be re-used. After completing the moth testing, clean the working platform, room floor and adjacent areas by mopping with 2 per cent bleaching powder solution. Maintain strict personnel hygiene.

Preparation of Permanent Slide of Pebrine Spores

☆ Make a thin smear on a glass slide and air dry.

- ☆ Fix in methyl alcohol for 2–3 minutes.
- ☆ Air-dry or blots dry the smear.
- ☆ Dilute giemsa stock solution (1 drop giemsa + 1 ml of distilled water) and stain the smear for 45–60 minutes.
- ☆ Wash in distilled water.
- ☆ Air dry or blot dry.
- ☆ Mount in DPX and examine with 600 magnification of microscope.
- ☆ In giemsa stain, pebrine spore appears as pinkish violet or bluish oval body.

Detection of Pebrine Spores in Soil/Dust

- ☆ Collect dust from the rearing house/egg production centers and equipments.
- ☆ Take 5 gm of the soil sample in a beaker and add 20 ml (1:4) of 0.6 per cent Potassium carbonate solution. Stir it thoroughly for 2 minutes. Allow the suspension to settle for 15–20 minutes.
- ☆ Centrifuge the supernatant for 5 minutes at 10,000 r.p.m.
- ☆ Collect the sediment.
- ☆ Take a drop of distilled water on a glass slide and a small quantity of sediment with a glass rod or brush.
- ☆ Add a tiny drop of Indian ink, mount and examine.

Detection of Pebrine Spores on Mulberry Leaves

- ☆ Wash 10–15 mulberry leaves in 100 ml of water.
- ☆ Centrifuge the wash for 8–10 minutes at 10,000 r.p.m.
- ☆ Discard the supernatant.
- ☆ Take a drop of distilled water on a glass slide and small quantity of the sediment with a glass rod or brush.
- ☆ Add a tiny drop of Indian ink, mount and examine.

Detection of pebrine Spores in Eggshell/Eggs/Un-hatched Eggs

- ☆ Collect head/body pigmented stage eggs/hatched eggshells/un-hatched eggs into a porcelain/glass mortar and add 0.6 per cent Potassium carbonate solution and grind thoroughly.

- ✩ After settling, filtering and centrifuging, examine the sediments after dissolving the same.

Detection of Pebrine Spores in Litters

- ✩ Collect 5 gm of litter in a mortar and add 40 ml 0.6 per cent potassium carbonate solution (8 times the weight of litter) and homogenize by grinding. If the resultant homogenate turns out to be viscous, add a few drops of HCl to reduce viscosity.
- ✩ After settling, filtering and centrifuging dissolve the sediment and observe under microscope.

Adoption of this method in stock race rearing, seed areas and also at silkworm seed rearing to detect spores will help to take up early remedial measures to contain the spread of the disease. The advantages are:

- ✩ Total elimination of wasteful sacrifice of larvae/pupae.
- ✩ Repeated faecal pellet examination may be conducted to ensure detection of infection in the silkworm population without additional cost.
- ✩ Random sampling method for faecal pellet collection instar-wise (Table 9.6) and centrifugation of homogenate sample enables accurate detection of spores to minimize the incidence of trans-ovarian transmission.

[B] Pests

Mulberry silkworm seed cocoons, pupae and sometimes ovipositing adult moths are attacked by several insect pests *viz.*, Coleoperans (*Lyprops cuticollis, Alphitobiulaevigatus, Necrobie rufipes, Tribolium castaeneum*) and many species of Dermestes (*Dermapteran labia arachidis)*; Tachinids (*Crossocosmia zebina* and *Ctenophorocera pavida)* and Acarid (mites) (*Pediculoids ventricosus)*. Of these, dermestid beetles and mites cause damage to a considerable level to grainages.

Table 9.6: Faecal Pellet Sample Size for 100 Dfls (after Patil *et al.*, 2001)

Age of Instars		*No. of Trays per Instars (4.5' diameter)*	*Sample Size*	
Instar	*Day*		*No. of Samples (10 gm each)*	*Total Weight of Faecal Matter (gm)*
I	2	2	1	10
II	2	2	1	10
III	2	4	2	20
IV	2	8	2	20
IV	4	10	3	30
V	2	18	4	40
V	4	25	6	60
V	6	30	8	80

(a) Dermestid Beetles

Dermestes belong to the order Coleoptera; super family Demestoidea and family Dermestidae. About 700 species of dermestid beetles are described. Five species of dermestid beetles *Dermestes alter, D. cadsvarinus, D. vulpinus, Attagenus fasiatus* and *Anthrenus verbesci* have been reported to attack the seed cocoons of mulberry silkworm (*Bombyx mori)* in the grainages (Sengupta *et al.*, 1990). They damage pupae and adults in the grainages in general and stored cocoons in particular.

Seasonal Occurrence

Presence of pest has been reported throughout the year and almost in all the sericultural zones of the country.

Biology

Adults of *Dermestes alter* are light yellow in colour which changes to dark brown or black and measures approximately 7 mm in body length. Female starts egg laying 5 days after emergence. The eggs are milky white, elongate and measures 1.90 x 0.48 mm in size. The eggs hatch in 5–7 days after oviposition. Newly hatched grubs are white in colour, 2.4 mm in length; spindle shaped and covered with hairs.

These grubs turn to black from second instars onwards. The grubs undergo 4–6 moults. The total life span of grub is 27–28 days after which it pupates. Pupal period is approximately 7–8 days after which emergence of adult takes place.

Dermestes cadverinus adults are dark brown in colour, 10 mm in length with oval elongated body and club shaped antennae. The eggs are oval, 2 mm in size and milky-white in colour, which hatches in about 7 days. Newly hatched grubs are spindle shaped reddish brown in color, covered with hairs. They start feeding on dried animal matter. The insect prefers to live in darker places. It moults 5–6 times in 4–8 weeks and become pupa. The larvae and adults are attracted by the smell of stifled cocoons and dried pupae inside. They bore into the cocoon and eat the dried pupae and thus damage cocoons making them unfit for further processing/reeling.

Adults of *Dermestes vulpinus* are shiny reddish brown to black in colour, sub-parallel in shape covered with dense hairs. This is a cosmopolitan species and is commonly known as hide beetle. The males are slightly smaller than females, bear a pit and a brush of hair on the fourth sternite and the four short basal tarsal segments of fore and mid legs lack the fine golden testaceous hair and also do not form ventral pads. Adult females lay eggs in batches of two or three, sometimes singly and oviposition is favoured under high humid conditions at 25–30°C temperature. The number of eggs laid by the adult varies from 200–850. Eggs are white in color and cylindrical in shape. Hatching takes place from the eggs after 2–4 days of oviposition. Depending on the temperature, humidity and kind of food, they moult 7–14 times and the larval period varies from 25–60 days. Full-grown larva is dark brown with a medium yellow stripe dorsally and is densely covered with hairs. They pupate in the last larval skin and the adult emerges from the pupa in 5–8 days. There may be 3–6 broods in a year depending on the temperature, humidity and other climatic conditions.

Symptoms and Nature of Damage

Grubs and adults make hole in the seed cocoon, enter inside and eat the pupae. They also damage pierced and melted cocoons. It is also observed that they sometimes attack ovipositing adults of silkworm in the grainages and the damage is caused mostly on the abdominal region.

Management

- Avoid longer storage of rejected/pierced cocoons and rejected layings.
- Collect the grubs and adults either by sweeping or using a vacuum cleaner and destroy them by burning or dipping in soap water so that continuity of life cycle of the pest is disrupted.
- Proper disinfections of grainage premises, grainage house and cocoon storage room besides periodical cleaning.
- Provide a wire mesh to the doors and windows of pierced cocoon storage rooms, grainage rooms to avoid entry of the beetles and grubs.
- Pierced cocoon storage room should invariable be constructed away from the grainage building.
- Dip wooden articles of the storage room and of the grainage building in 0.2 per cent Malathion solution for 2-3 minutes. Safe period for the use of appliances is 10 days after thoroughly washing with water and sun drying.
- Dust 5 per cent malathion on pierced cocoons before storage. Approximately 5 kg dust is sufficient to cover 50 kg of pierced cocoons.
- Fumigate dried cocoon (not live) storage room with Methyl bromide at 0.5 gm per $3m^2$ for 3 days.
- Store pierced cocoons in deltamethrin treated bags by dipping them in 0.028 per cent deltamethrin solution (1 part deltamethrin–2.8 per cent EC in 100 parts of water) followed by drying in shade.
- Spray 0.028 per cent deltamethrin solution on walls and floor of PC room once in three months.
- Dust commercial grade bleaching powder (30 per cent chlorine) at the rate of 200 gm/m^2 all around inner wall of pierced cocoon room to prevent crawling of grubs from PC room.

(b) Mites

Pediculoides ventricosus (Newport) commonly known as 'straw itch' mite belongs to the order Acarina and family Pediculoididae.

Table 9.7: A Tentative World list of Dermestid Insect Pests of Stored Silkworm Cocoons and their Products (Vijay Veer *et al.*, 1996)

Species	*Commodity Damaged*	*Country*
Thylodrius contractus	Silk fabrics	USA
Dermestes alter	Cocoons, Raw silk	Germany, India, USA, Japan
Dermestes coarctatus	Eggs, Cocoons, Silk	Japan
Dermestes frischii	Cocoons, raw silk	Italy, Central Asia
Dermestes lardarius	Cocoons, Eggs, Caterpillars	Bulgaria, France, Italy
Dermestes leechi	Cocoons	India
Dermestes maculates	Moth, Raw silk	Italy, India
Dermestes murinus	Cocoons, Raw silk	Italy
Dermestes peruvianus	Raw silk	Britain
Dermestes tesclataocollis	Raw silk	India
Dermestes undulates	Cocoons, Raw silk	Central Asia, India
Dermestes vorax	Cocoons, Raw silk	India
Attagenus fasciatus	Cocoons, Fabrics	India
Attagenus birmanicus	Cocoons	India
Attagenus pellio	Cocoons	Germany, Georgia
Attagenus unicolor	Cocoons, Fabrics	USSR, N. America
Attagenus unicolor japonicus	Cocoons, Raw silk, Floss, Fabrics	Japan
Trogoderma halsteadi	Cocoons	India
Trogoderma sternale	Cocoons, Raw silk	USA
Trogoderma varium	Cocoons, Eggs, Raw silk	Japan
Trogoderma versicolor	Cocoons, Raw silk	Georgia, Central Asia
Orphinus fulvipes	Cocoons	India
Anthrenus flavipes	Cocoons, Fabrics	Africa, India, Germany
Anthrenus museorum	Silk	Germany
Anthrenus pimpinellae	Cocoons	India
Anthrenus scrophularae	Silk	North America
Anthrenus verbasci	Cocoons, Raw silk, Fabrics, Eggs	Canada, Italy, Japan, USA
Trnodes rufesceens	Cocoons, Raw silk	Japan

Seasonal Occurrence

Its occurrence has been reported throughout the country mostly from May to September.

Biology

The newly born mite is 0.2 mm in length and light yellow in colour. Males hatch earlier than females. Females are spindle shaped while males' oval shaped. Head is triangular. They have four pairs of legs with small claws. They are ovoviviparous. Females after mating and getting a suitable substratum attach themselves with claws and suckers. The young ones hatch out from the eggs in the female body and pass out in the form of adult like small acarid. Each adult female produces 100–150 young ones. There are 17 generations in a year and time taken to complete each generation is in 7–18 days.

Symptoms and Nature of Damage

The acarid attacks larvae, pupae and adults moths of the silkworm. Acarid attaches itself to the soft skin of the host between the segments and obtains nutrition from there. They also produce a kind of toxin, which ultimately kills the host silkworm. After being infested, the larva stops feeding and become sluggish. The body of the host turns purple brown and silkworm starts vomiting yellowish brown fluid. The worms die within one or two days after infection.

Management

- ☆ On seeing the attack of acarid, grainage trays should be replaced immediately.
- ☆ Disinfect the grainage appliances with steam.
- ☆ Wheat and rice straw or cotton should not be stored near the grainage building.

Chapter 10
RECORDS

Grainages (silkworm egg producing centers) are required to maintain record for certain specific information pertaining to seed rearers, crop supervision, larval examination, purchase of seed cocoons, available seed cocoon stock at a given date, cocoon assessment, pupa/moth examination report, pairing and isolation details, daily egg card preparation details and daily egg card abstract, Dfls preparation, seed cocoon stock cum preparation details, egg sheet account, pierced cocoon stock, records of temperature and humidity, Dfls indent and supply, Dfls day-to-day stock, cash bill, invoice, remittance, abstract of daily remittance, disposal of Dfls, DCB, consignment and release of laying, harvest details and test hatching records. All the above records are to be maintained in register and should be made available for audit and for inspection on visit of any responsible officer. From time to time the formats to be utilized to maintain various records are modified, improved and recommended by different workers. The formats which are presently in use is compiled and published by Jayaswal *et al.* (2008). The details of the formats are furnished below (Table 10.1 to Table 10.26).

A. PRODUCTION

Table 10.1: Adopted Seed Rearers List

Sl.No.	*Name of the Rearers*	*Village/ District*	*Area Under Mulberry (Acre)*	*Rearing Capacity (Dfls/crop)*

Table 10.2: Checklist for the Seed Rearers

a. Name of the farmer and address:

b. Experience in sericulture (total years):

c. Details of mulberry garden

- ☆ Variety and spacing:
- ☆ Area under mulberry:
- ☆ Irrigation facility:
- ☆ Manuring schedule and doses:
- ☆ Fertilizer schedule and doses:
- ☆ Pruning schedule:

d. Disinfection (Formalin/Bleaching powder/Serichlor/Asthra etc – specify):

e. Methods followed for disinfections (sprayer/fumigation etc – specify):

f. Bed disinfectant used (Vijetha/Resham Jyothi/Labex etc – specify):

g. Types of rearing house (RCC/Mud/Thatched etc – specify):

h. Types of rearing (Shelf rearing/Tray rearing/box rearing/floor rearing etc):

i. System of rearing (Shoot rearing/leaf rearing etc.):

j. Number of crops/year:

k. Number of Dfls/crop:

l. Types and No. of mountages used (Plastic/bamboo/rotary etc.):

m. Whether the farmer meets the terms and conditions of the grainage:

n. Whether the farmer can be adopted?

Table 10.3: Crop Supervision Report

Crop Number: **Rearers details:**

Sl.No.	Particulars
1.	Rearer's name
2.	Address
3.	Date of disinfection
4.	Breed
5.	Quantity of eggs distributed
6.	Source of layings
7.	Lot number
8.	Date of release of Dfls
9.	Hatching per cent and date of hatching
10.	Fecundity (Nos.)
11.	Chawki centre name
12.	Date of supply of Chawki larvae
13.	Spinning date of larvae
14.	Cocoon harvesting date
15.	Yield/100 Dfls (kgs)
16.	Single cocoon and shell weight (gm)
17.	Cocoon marketing date
15.	Pupation (per cent)
16.	Certificate for disease freeness

Crop Supervision Summary

Sl.No.	Supervision/ Examination Date	Examination Report	Suggestions Given	Name of Officer	Signature

Table 10.4: Larval Examination Report

Sl.No.	Date	Name and Address of Rearer	Breed	Age/ Stage of Larvae	Result	Tested by	Remarks

Table 10.5: Purchase of Seed Cocoons

Sl.No.	Name and Address of rearer	Breed	Spun on Date	Quantity		Rate/kg Cocoons	Amount (Rs.)	Other Charge (Rs.)	Total Amount (Rs.)	Number of Good Cocoons/kg
				By Number	By Weight					

Quantity of Cocoons Purchased				Remarks
Actual Cocoons		Good Cocoons		
By Number	By Weight	By Number	By Weight	
				Number of cocoons/kg: Number of good cocoons/kg: Single cocoon weight (gm): Single shell weight (gm): Shell ratio (per cent): Yield/100 Dfls (kgs.): Pupation rate (per cent):

Table 10.6: Seed Cocoon Stock Details

Sl.No.	Lot No.	Name and Address of Rearer	Invoice Number and Date	Breed	Source	Spun on Date	Purchase on Date

Received on Date	Cocoon Numbers/ kg	Quantity of Cocoons Purchase				Rate (Rs.)
		Actual Cocoons		Good Cocoons		
		By wt.	By No.	By wt.	By No.	

Total Amount (Rs.)	Market Fees (Rs.)	Grand Total (Rs.)	Payment Details	Signature of In-charge

Table 10.7: Cocoon Assessment Details

Sl. No.	Lot Number	Breed	Source	Quantity purchased		Cocoons Number/ kg.	Quantity Taken for Assessment	
				By Weight	By Number		By Weight	By Number

Spun on	Assessment on Date	Rejection								Total Rejection	
		Melt		Uzi		Muscardine		Flimsy			
		No.	%	No.	%	No.	%	No.	%	No.	%

Good Cocoons per kg.		Male		Female		Single Cocoon Weight (gm)	Single Shell Weight (gm)	Shell Ratio (%)
No.	%	No.	%	No.	%			

Table 10.8: Pupa/Moth Examination Details

Sl.No.	Lot Number	Date of Testing	Breed/ Hybrid	Pupa/ moth Tested (No.)	Laid on Date	Loose/ Sheet Eggs	No. of Sheets	No. of Pairs	Sheets tested	
									Book No.	Sheet No.

No. of Samples Tested	No. of Samples Revealed Pebrine	Pebrine %	Day Wise % of Testing	Examined by	Cross Examined by	Remarks

Table 10.9: Pairing and Isolation Details

Lot Number:		Quantity of Cocoons:		Spun on:		Source:
Date of Emergence	Combination	Book Number	Sheet Number	Number of Sheets	Number of Pairs	Remarks

Table 10.10: Egg Card Details

Lot number:		Combination:	
Laid on:		Accounted on:	
Book Number	Sheet Number	Rejection per Sheet	Dfls Obtained per Sheet

Egg Card Abstract

Lot No.:		Quantity of Cocoons:		Spun on:		Source:		
Sl.No.	Date of Emergence	Tested on	No. of Pairs	No. of Rejections	No. of Dfls	% of Pairs	% of Rejections	% of Dfls

Table 10.11: Egg Card Details for Hybrid

Sl.No.	Lot No.	Laid on Date	Accounted on Date	Combination	Number of Sheets	Number of Pairs	Number of Dfls	Weight (gm)

Table 10.12: Dfls Preparation Details

Sl.No.	Lot No.	No. of Cocoons	Spun on Date	Source	Emergence Date	Combination	Book Number	Sheet Number

No. of Sheets	Pairs		Rejection		Dfls		Weight of Loose Eggs (kg)	Egg Yield/ kg of Cocoons (gm)
	No.	%	No.	%	No.	%		

Table 10.13: Seed Cocoon Stock-cum-Preparation Details

Sl.No.	Lot No.	Name and Address of Rearer	Breed	Source	Spun on	Marketed on	Supplied on	Cocoons/kg	
								Actual	Good

Quantity of Cocoons Purchased				Rate/ kg	Amount (Rs.)	Other Charges (Rs.)	Total Amount (Rs.)	Payment Details
Actual Cocoons		Good Cocoons						
By Wt.	By No.	By Wt.	By No.					

Total Rejection Plus Male or Female Cocoons Marketed		Amount Realized (Rs.)	Actual Cost of Seed Cocoons (Rs.)	Date of Emergence	Combination	Book Number
No.	kg.					

Sheet No.	No. of Sheets	Total Pairs	Total Rejection	Quantity of Dfls		Wt. of Loose Eggs (kg)	Total Dfls	Per cent of			Egg Yield/ kg of Cocoons
				Treated	Hibernated			Pairs	Rejection	Dfls	

Pierced Cocoons							Remarks
Quantity of PC (kg)	No. of PC/kg	Quantity of PC (kg)		Date of Lifting	Quantity Lifted	Stock on Hand	
		D.M.	U.E.M.				

Table 10.14: Egg Sheet Account Details

Date	Opening Balance	Invoice Number and Date	Book Number	Sheet Number	Quantity Received	Total

Book Number	Sheet Number	Number of Sheets	Lot Number	Closing Balance	Signature	Remarks

Table 10.15: Pierced Cocoon Stock Details

Sl.No.	Lot Number	Breed	Actual Quantity (Number)	Number of Cocoons/kg (when received)

Pierced Cocoons							Remarks
Quantity of PC (kg)	No. of PC/ kg	Quantity of PC (kg)		Date of Lifting	Quantity Lifted	Stock on Hand	
		D.M.	U.E.M.				

Table 10.16: Records of Temperature and Humidity for the Month of ———

Date	6 A.M.			11 A.M.		
	Dry °C	Wet °C	RH %	Dry °C	Wet °C	RH%

2 P.M.			5 P.M.			Remarks
Dry °C	Wet °C	RH %	Dry °C	Wet °C	RH %	

Abstract for the month of______

Particulars		Maximum	Minimum	Average
Average temperature	Dry			
	Wet			
Relative Humidity (per cent)				

B. MARKETING

Table 10.17: Dfls Indent and Supply Details

Sl.No.	Date of Indent	Name, Address and Phone Number of Indenter	Mode of Indent			Quantity Indented		
			Letter	Phone	Person	Bi x Bi	M x Bi	Others

Release Date	Hatching Date	Lot No.	Supply Date	Sale Details		Quantity of Dfls Supplied		
				Cash Bill No. and Date	Invoice No. and Date	Bi x Bi	M x Bi	Others

Table 10.18: Dfls Day-to-Day Stock Details

Sl.No.	Date	Breed/ Combination	Opening Balance	Preparation of the Day/ season	Dfls Received	
					From Other Centers	From Cold Storage

Total	Dfls Distribution				Closing Balance
	Sent to Other Centers by Invoice	Sent to Cold Storage	Sold on the Day by Cash	Recommendation to Write-off	

Table 10.19: Cash bill

Sri________________ Cash bill number________________

________________ Date________________

Sl.No.	Particulars			Quantity	Rate		Amount	
	Combination	Book No.	Sheet No.		Rs.	Ps.	Rs.	Ps.

Rupees in words________________________________

The Dfls received in good condition

Signature of the rearer/receiver________________ Signature of the official

Table 10.20: Invoice

To,

Invoice number________________

Date________________

As per indent No.________________Dated________________following quantities of Dfls/seed cocoons are sent through________________. Kindly acknowledge by returning the duplicate copy duly stock certified and signed.

Sl. No.	Lot No.	Breed/ Hybrid	Book No.	Egg Sheet No.	Quantity	Laid on	To Hatch on	Rate/ 100 Dfls	Total Amount (Rs.)

Rupees in words________________________________

Receiver:

Signature of Disposal I/c

Name:

Designation:

Place: Signature of Grainage I/c

Table 10.21: Remittance Details

Date of sales: Date of remittance:

Cash Bill No. and Date (cash Receipt)	Lot No.	Cash Sales			Credit Sales			
		No. of Dfls	Rate/ 100 Dfls	Amount (Rs.)	Invoice No. and Date	No. of Dfls	Rate/ 100 Dfls	Amount (Rs.)

Abstract of Daily Remittance

Breed/ Combination	Quantity of Dfls		Amount		Total Amount
	Cash	Credit	Cash	Credit	

Table 10.22: Disposal Details

Lot number: Total quantity of Dfls prepared:

Combination:

Sl.No.	Name and Address of the Rearer	Book Number	Sheet Number	Quantity of Dfls	Rate/100 Dfls (Rs.)

Cash Sales			Credit Sales			Realization		
Bill No.	Date	Amount	Invoice No.	Date	Amount	Bill No.	Date	Amount

Table 10.23: DCB Details

Name and Designation:

Sl.No.	Invoice No.	Date	Lot No.	Combination	Quantity of Dfls	Rate/ 100 Dfls (Rs.)	Amount (Rs.)	Full/ Part Payment

Quantity	Rate/ 100 Dfls	Amount (Rs.)	Cash Receipt No. and Date	Balance Amount Outstanding	Signature of Incharge	Remarks

Table 10.24: Consignment and Release Details

Sl. No.	Lot No.	Hybrid	Laid on Date	No. of Dfls		Consigned on Date	Schedule	Consigned by	Due Release Date
				By Wt.	By No.				

Quantity of Dfls Released		Date of Release	Released by	No. of Dfls Rejected	Balance Quantity of Dfls	Initials	Remarks
By Wt.	By No.						

Table 10.25: Harvest Details

Sl.No.	Rearers' Name	Address	Lot No.	Hybrid	Dfls Quantity	Total Yield (kg)	Yield/100 Dfls (kg)	Rate/ kg (Rs.)	Total Amount (Rs.)	Remarks

Table 10.26: Test Hatching Details

Sl.No.	Lot No.	Source	Laid on Date	Released on Date	Expected Date of Hatching

Particulars					Hatching (per cent)
Unfertilized Eggs	Unhatched Eggs (Blue eggs)	Dead Eggs	Hatched Larvae	Total Eggs (Fecundity)	

BIBLIOGRAPHY

Abe, Y. and Fujiwara, T. (1979) Mode of multiplication of protozoan, *Pleistophora* sp. (Microsporidia–Nosematidae) in the midgut epithelium of the silkworm larvae. *J. Seric. Sci. Japan*, 48: 19-23.

Akai, H. (1958) Observations of the micropylar region of the chorion in the silkworm, *Bombyx mori* L. *Acta Sericol.*, 26: 19-21.

Akutsu, S. and Yoshitake, N. (1977) Electron microscopic observation on the vitelline membrane formation in the silkworm, *Bombyx mori* L. *J. Seric. Sci. Japan*, 46: 509–514.

Alenkseenork, A. (1986) Nosematol for control of nosematosis of bees and pebrine of silkworm. *Veterinariya, USSR*, 10: 45-47.

Ananthalakshmi K.V.V.; Fujiwara, T. and Datta, R.K. (1994) First report on the isolation of three microsporidians (*Nosema* spp.) from the silkworm, *Bombyx mori* L. in India. *Indian J. Seric.*, 33: 146-148.

Ando, Y. (1974) Egg diapause and ambient oxygen tension in the false melon beetle, *Atrachya menethesi* (Coleoptera: Chrysomelidae). *Appl. Ent. Zool.*, 9: 261-270.

Anonymous (1980) *Text Book of Tropical Sericulture*. Japan Overseas Cooperation Volunteers, Tokyo, Japan, pp. 551-556.

Askari, S. and Sharan, R.K. (1984) Studies on different copulating duration on pre-oviposition, fecundity and fertility of mulberry silkworm, *Bombyx mori* L. (Bombycidae: Lepidoptera). *J. Adv. Zool.*, 5: 114-119.

Austrurov, B.L.; Baburashvli, E.L.; Bedniakova, T.A.; Verciskaia, V.N.; Labzhanidze V.I. and Ovanesyan T.T. (1969) Thermal intravital disinfections of eggs as a new method of control of silkworm pebrine disease. *Zvakad Nauk, SSSR, Ser. Bio.*, 6: 811-818.

Ayuzawa, C.I.; Sekido, K.; Yamakawa, U.; Sakurai, W.; Kurata, Y.; Yaginuma, Y. and Tolkoro, Y. (1972) *Handbook of Silkworm Rearing. Agric. Tech. Manual.* Fuji Publication Co. Tokyo, Japan, pp. 114-135.

Azuma, M. and Yamashita, O. (1985) Immunohistochemical and biochemical localization of trehalase in the developing ovaries of the silkworm, *Bombyx mori. Insect Biochem.*, 15: 589-596.

Baig, M.; Datta, R.K.; Nataraju, B.; Samson, M.V. and Sivaprasad, V. (1992a) Protein-A linked latex antiserum test for the detection of *Nosema bombycis* (Nageli) spores. *J. Invertebr. Patho.*, 60: 310-313.

Baig, M.; Gupta, S.K.; Nataraju, B.; Mohd. Sha, M.M.; Sivaprasad, V.; Datta, R.K.; Gupta, R. and Samson, M.V. (1992b) Latex agglutination test for the detection of pebrine in the silkworm, *Bombyx mori* L. *Indian J. Seric.*, 31: 141-145.

Baig, M.; Samson, M.V.; Sasidharan, T.O.; Sharma, S.D.; Balavenkatasubbaiah, M. and Jolly, M.S. (1988a) Study on the spread of pebrine after introduction of transovarially infected worms in a colony of silkworm, *Bombyx mori* L. *Sericologia*, 28: 75-80.

Baig, M.; Samson, M.V.; Sharma, S.D.; Balavenkatasubbaiah, M.; Sasidharan, T.O. and Jolly, M.S. (1988b) Effect of certain disinfectants as surface sterilants against pebrine in surface contaminated laying. *Sericologia*, 28: 81-87.

Bakke, A. (1969) Extremely low super-cooling point in eggs of *Zeiraphera diniara* (Guenec) (Lepidoptera: Tortricidae). *Norsk. Ent. Tidsskr.*, 16: 81-83.

Bale, J.S. (2002) Insects and low temperatures: from Molecular Biology to Distributions and Abundance. *Philosophical Transactions of the Royal Society of London. Biological Sciences*, 357: 849-862.

Balinsky, B.I. (1981) *An Introduction to Embryology*–Saunders College, Philadelphia.

Bedniakova, T.A. and Vereiskava, V.N. (1958) The disinfection action of high temperature on eggs on mulberry silkworm (*B. mori* L) infected with pebrine (*N. bombycis*) at different stages of the diapausal cycle of development (in Russian). *Doklady Akad Nauk, USSR, Biol Sci Sect.*, 122: 760-763.

Behrena W (1985) Environmental aspects of insect dormancy; in *Environmental Physiology and Biochemistry of Insects*. Hoffmann, K.H. (ed.), pp. 68-94, Springer Verlag, Berlin, Tokyo.

Benchamin, K.V.; Holkar, P.S. and Rao, V. (1990) Studies on the separation of fertile and unfertile eggs in silkworm. *Indian J. Eco.*, 17: 79-80.

Benchamin, K.V.; Magdum, S.B. and Shivashankar, N. (1990) Mating capacity of male moths of pure breeds and hybrids in silkworm, *Bombyx mori* L. *Indian J. Seric.*, 29: 182-187.

Bennett, V.A.; Pruitt, N.L. and Lee R.E, Jr. (1997) Seasonal changes in fatty acid composition associated with cold-hardening in 3rd instar larvae of *Eurosta solidaginis*. *J. Comp. Physiol. B. Biochem., Syste. & Environ. Physiol.*, 167: 249-255.

Bennett, V.A.; Sformo, T.; Walters, K.; Toien, O.; Jeannet, K.; Hochstrasser, R.; Pan, Q.; Serianni, A.S.; Barnes, B.M. and Duman, J.G. (2005) Comparative over-wintering physiology of Alaska and Indiana populations of the beetle *Cucujus clavipes* (Fabricus): Role of antifreeze proteins, polyols, dehydration and diapause. *J. Experimental Biol.*, 208: 4467-4477.

Biram Saheb, N.M.; Gaur, J.P.; Samson, M.V. and Rajanna, K.L. (1998) Behaviour of male moth in *Bombyx mori* L. *Indian Text. J.*, 108: 46-49.

Biram Saheb, N.M. and Gowda, P. (1987) Silkworm Seed Technology; in *Appropriate Sericulture Techniques*. Published by Central Sericultural Research &Training Institute, Mysore.

Biram Saheb, N.M.; Sengupta, K. and Vemananda Reddy, G. (1990) *A Treatise of the Acid Treatment of Silkworm Eggs*. CSR&TI, Mysore, Central Silk Board, India.

Biram Saheb, N.M.; Singh, T. and Saratchandra, B. (2009) Occurrence of unfertilized eggs in the mulberry silkworm, *Bombyx mori* (L.) (Lepidoptera: Bombycidae). *Intl. J. Indust. Entomol.* (Korea), 18: 1-7.

Biram Saheb, N.M.; Singh, T.; Kalappa, H.K. and Saratchandra, B. (2005) Mating behaviour in mulberry silkworm, *Bombyx mori* L. *Intl. J. Indust. Entomol.* (Korea), 10: 87-94.

Biram Saheb, N.M.; Vinod, K.; Negi, B.B.S.; Sengupta, K. and Noamani, M.K.R. (1991) Comparative study on age-specific responses of hot and cold hydrochlorization methods on the diapausing eggs of silkworm, *Bombyx mori* L. *Indian J. Seric.*, 30: 151-153.

Biram Saheb, N.M.; Vinod, K.; Negi, B.B.S. and Samson, M.V. (1996) Acid treatment in relation to refrigeration of silkworm, *Bombyx mori* eggs. *Indian J. Seric.*, 35: 77-79.

Block, W. (2002) Interactions of water, ice nucleators and desiccation in invertebrate cold survival. *European J. Entomol.*, 99: 259-266.

Bliss, C.I. (1927) The oviposition rate of the grape leafhoppers. *J. Agric. Res.*, 34: 847-852.

Block, W. and Zettel, J. (2003) Activity and dormancy in relation to body water and cold tolerance in a winter-active springtail (Collembola). *European J. Entomol.*, 100: 305-312.

Bottrell, H.H. (1975) The relationship between temperature and duration of egg development in some epiphytic Cladocera and Copepoda from the river Thames–reading with a discussion of temperature functions. *Oecologia*, 18: 63-85.

Brady, U.E. (1983) Prostaglandin in insects. *Insect Biochem.*, 13: 443-451.

Braune, H.J. (1976) Effect of temperature on the rate of oxygen consumption during morphogenesis and diapause in the egg stage of *Leptopterna dolobrata* (Heteroptera: Miridae). *Oecologia*, 19: 77-87.

Brower, L.P. (1995) Understanding and misunderstanding the migration of the Monarch butterfly (Nymphalidae) in North America: 1857-1995. *J. Lepidopterists Society*, 49: 304-385.

Canning, E.U. (1993) *Parasitic Protozoa*. Kreier, J.P. and Baker, J.R. (eds.), Academic Press, London, 6: 299-370.

Canning, E.U.; Curry, A.; Cheney, S.A.; Lafranchi-Tristem, N.J.; Kawakami, Y.; Hatakeyama, Y.; Iwano, H. and Ishihara, R. (1999) *Nosema tyriae* and *Nosema* sp., microsporidian parasites of Cinnabar moth *Tyria jacobaeae*. *J. Invertebr. Pathol.*, 74: 29-38

Carlos, A.; Didier, .P and Jean, C. (1996) Recovery and characterization of a replicas complex in rotavirus-infected cells by using a monoclonal antibody against NSP2. *J. Virol.*, 70: 985-991.

Chandra, A.K. and Sahakundu, A.K. (1983) The effect of drug on pebrine infection in *Bombyx mori* L. *Indian J. Seric.*, 21 & 22: 67-69.

Chattopadhyay, S. (1995) Mating duration and production of viable eggs of mulberry silkworm (*Bombyx mori* L). *Environ. Ecol.*, 13: 460-461.

Chaturvedi, M.L. and Upadhyay, V.B. (1990) Effect of cold storage on the hatchability of silkworm (*Bombyx mori*) eggs. *J. Adv. Zool.*, 11: 63-65.

Chen, P.S. (1971) *Biochemical Aspects of Insect Development*. S. Karger Basel, Switzerland.

Chen, J.; Teng, J.; Hu, C. and Michael, M. (1989) Production of monoclonal antibodies to densovirus of silkworm (*Bombyx mori*) and their application in diagnosis. *Chinese J. Virol.*, 5: 77-81.

Chen, Q.; Tao, T.; Li, S. and Sheng, J. (1994) Studies on the three-step refrigeration system of the hibernating silkworm eggs. *Bull. Seric. (China)*, 25: 11-14.

Cheung, W.W.K. and Wang, J.B. (1995) Electron microscopic studies on *Nosema mesnili* (Microsporidia: Nosematidae) infecting the malpighian tubules of *Pieris canidia* larva. *Protoplasma*, 186: 142-148.

Chino, H. (1957a) Carbohydrate metabolism in diapause eggs of the silkworm, *Bombyx mori*. I. Diapause and change of glycogen content. *Embryologia*, 3: 295–316.

Chino, H. (1957b) Conversion of glycogen to sorbitol and glycerol in the diapause eggs of the *Bombyx mori* silkworm. *Nature*, 180: 606 -607.

Chino, H. (1958) Carbohydrate metabolism in diapause eggs of the silkworm, *Bombyx mori*. II. Conversion of glycogen into sorbitol and glycerol during diapause. *J. Insect Physiol.*, 2: 1–12.

Chitra, C.; Karanth, N.G.K. and Vasantharajan, V.N. (1975) Diseases of the mulberry silkworm, *Bombyx mori* L. *J. Sci. Indust. Res.*, 34: 386-401.

Christiana, S.T. and Kamble, C.K. (1995) Loose egg preparation and adoption–an appraisal. *J. Seric.*, 32: 40-45.

Christiana, S.T.; Jayaramaraju, P.; Vijayakumar, S.; Manjula, A. and Veeraiah, T.M. (2003) Studies on the effect of cold storage of seed cocoons on the occurrence of unfertilized eggs in silkworm, *Bombyx mori* L. Natl. Semi. Silkworm Seed Production, Silkworm Seed Technology Laboratory, Bangalore, India, pp. 145-148.

Chowdhary, S.N. (1967) *The Silkworm and its Culture*. Mysore Printing and Publishing House, Mysore, India, pp. 76.

Coulon, M. (1988) Comparative changes of ecdysteroid content in *Bombyx mori* eggs in diapausing and non-diapausing development. *Comp. Biochem. Physiol.*, 89: 503-509.

Craig, E.A.; Gambill, B.D. and Nelson, R.J. (1993) Heat shock proteins: molecular aspects of protein biogenesis. *Microbiological Reviews*, 57: 402-414.

Danks, H.V. (2000) Measuring and reporting life-cycle duration in insects and arachnids. *European J. Entomol.*, 97: 285-303.

Danks, H.V. (2001) The nature of dormancy responses in insects. *Acta Societatis Zoologicae Bohemicae*, 65: 169-179.

Danks, H.V. (2005) Key themes in the study of seasonal adaptations in insects. I. Patterns of cold hardiness. *Applied Entomol. Zool.*, 40: 199-211.

Danks, H.V. (2006) Insect adaptations to cold and changing environments. *Can. Entomol.*, 138: 1-23.

Danks, H.V. (2007) The elements of seasonal adaptations in insects. *Can. Entomol.*, 139: 1-44.

Das, S.; Saha, A.K. and Shamsuddin, M. (1996) High temperature induced sterility in silkworm. *Indian Silk*, 35: 26-28.

Datta, R.K. (1988) Embryological studies on non-diapause silkworm, *Bombyx mori* L. *Indian J. Seric.*, 27: 1-6.

Datta, R.K.; Sengupta, K. and Biswas, S.N. (1972) Studies on the preservation of multivoltine silkworm eggs at low temperatures. *Indian J. Seric.*, 11: 20-27.

Dautel, H. (1999) Water loss and metabolic water in starving *Argas reflexus* nymphs (Acari: Argasidae). *J. Insect Physiol.*, 45: 55-63.

Davey, K.G. (1985) The female reproductive tract; in *Comprehensive Insect Physiology, Biochemistry and Pharmacology*. Kerkut, G.A. and Gilbert, L.I. (eds.), Vol. I., pp 15-36, Pregamon Press, Oxford.

Denlinger, D.L. (1985) Hormonal control of diapause; in *Comprehensive Insect Physiology, Biochemistry and Pharmacology*. Kerkut, G.A. and Gilbert, L.I. (eds.), Vol. VIII. pp. 353-421, Pregamon Press, Oxford.

Denlinger, D.L. (1991) Relationship between cold hardiness and diapause; in *Insects at Low Temperature*. Lee, R.E. Jr. and Denlinger, D.L. (eds.), pp 174-198, Chapman and Hall, New York.

Devaiah, M.C. and Krishnaswami, S. (1975) Observations on the seasonal incidence of pebrine disease of the silkworm, *Bombyx mori* L. *Indian J. Seric.*, 14: 27-30.

Duman, J.G.; Wu, D.W.; Xu, L.; Turstman, D. and Olsen, T.M. (1991) Adaptations of insects to subzero temperatures. *Rev. Biol.*, 66: 387-410.

Ehrlich, P.R. and Raven, P.H. (1964) Butterflies and plants: a study in co-evolution. *Evolution*, 18: 586-608.

Fu, W. and He, M. (1993) Relationship between the maturity of mother moths and amount of laid eggs. *Bull. Seric.* (China), 24: 29-30.

Fugo, H. and Arisawa, N. (1992) Oviposition behaviour of the silkmoths, which mated with males, sterilized by high temperature in the silkworm, *Bombyx mori*. *J. Seric. Sci. Japan*, 61: 110-115.

Fujimoto, N. (1951) Laying of eggs on the inclined plane in the silkworm moth. *J. Seric. Sci. Japan*, 20: 258.

Fujiwara, T. (1979) Infectivity and pathogenecity of *Nosema bombycis* to larvae of the silkworm. *J. Seric. Sci. Japan*, 48: 376-380.

Fujiwara, T. (1980) Three Microsporidians (*Nosema* spp.) from silkworm *Bombyx mori*. *J. Seric. Sci. Japan*, 49: 229-236.

Fujiwara, T. (1984a) A *Pleistophora* – like microsporidian isolated from the silkworm, *Bombyx mori. J. Seric. Sci. Japan*, 53: 398-402.

Fujiwara, T. (1984b) *Thelohania* sp. (Microsporidia – Thelohanidae) isolated from the silkworm, *Bombyx mori. J. Seric. Sci. Japan*, 53: 459-460.

Fujiwara, T. (1985) Microsporidia from silk moths in egg production sericulture. *J. Seric. Sci. Japan*, 54: 108-111.

Fujiwara, T. (1993) Comprehensive report on the silkworm disease control. A report on "*Bivoltine Sericulture Technology Development in India*" submitted to Central Silk Board, Bangalore, India, pp.110.

Fujiwara, T. and Kagawa, T. (1984) Control of *Nosema bombycis* parasitizing silkworm eggs by treatment with hydrochloric acid on exposure to various temperatures. *J. Seric. Sci. Japan*, 53: 394-397.

Fukuda, S. and Tekeuchi, S. (1967) Diapause–factor producing cells in the suboesophageal ganglion of the silkworm, *Bombyx mori* L. *Proc. Jap. Acad.*, 43: 51-56.

Furasawa, T. and Shikata, M. (1982) Temperature dependent changes of polyols and glycogen content in the eggs of silkworm, *Bombyx mori. J. Seric. Sci. Japan*, 51:77-83.

Furasawa, T.; Shikata, M. and Yamashita, O. (1982) Temperature dependent sorbitol utilization in diapause eggs of the silkworm, *Bombyx mori. J. Comp. Physiol.*, 157: 21-26.

Furasawa, T.; Shimizu, K. and Yano, T. (1987) Polyol accumulation in non-diapause eggs of the silkworm, *Bombyx mori. J. Seric. Sci. Japan*, 56:150-156.

Furasawa, T.; Yaginuma, T. and Yamashita, O (1992) Temperature–induced metabolic shifts in diapause and non-diapause eggs of the silkworm, *Bombyx mori. Zool. Jb. Physiol.*, 96: 169-178.

Furasawa, T. and Yang, Won-jin (1987) Fluctuations of free sugars with embryonic developments of non-diapause eggs of the silkworm, *Bombyx mori. J. Seric. Sci. Japan*, 56: 143-149.

Geetha Bai, M.; Patil, C.S. and Kasturi Bai, A.R. (1985) A new method for easy detection of pebrine spores. *Sericologia*, 25: 297-300.

Genkai, K.M.; Mitsuhashi, H.; Kohmatsu, Y.; Miyaska, H.; Nozaki, K. and Nakanishi, M. (2005) A seasonal change in the distribution of a stream-dwelling stonefly nymph reflects oxygen supply and water flow. *Ecological Research*, 20: 223-226.

Giebutawicz, J.M.; Raina, R.A. and Uebel, *E.G.* (1990) Mated like behaviour in senescent virgin females of Gypsy moth, *Lymantria. J. Insect Physiol.*, 36: 485-498.

Gochnaner, A. and Margetts, P. (1980) A microsporidian disease in the silkworm, *Bombyx mori* L. *Sericologia*, 20: 201-210.

Goldsmith, R.; Paule, M.; Weare, B. and Clermont-Rattner, E. (1978) Development of the chorion in *Bombyx mori. J. Cell Biol.*, 79: 32.

Govindan, R.; Narayanaswamy, T.K. and Devaiah, M.C. (1998) *Principles of Silkworm Pathology*. Seri Scientific Publishers, Bangalore, India, pp.420.

Gowda, B.L.V.; Sannaveerappanayar, V.T. and Shivayogeshwar, B. (1989) Fecundity and hatchability in mulberry silkworm, *Bombyx mori* (L) as influenced by pupal weight. *Intl. Cong. Trop. Seric. Pract.* Bangalore, Feb. 18-23, Part VI: 21-24.

Gowda, P. (1988) Studies on some aspects of egg production in silkworm, *Bombyx mori* L. Ph.D. Thesis, Univ. Mysore, India.

Gowda, P. and Jolly, M.S. (1987) Acid treatment of silkworm eggs made easy. *Indian Silk*, 26: 33-36.

Grobov, O.F. and Rodionova, Z.E. (1985) Identification of spores of *Nosema bombycis* from silkworm. *Veterinarira, Moscow USSR*, 12: 70-71.

Guerine-Menevillae (1849) cit. in; *Pebrine Disease of Silkworm*–a technical report.

Tatsuke, K. (ed.) (1971), Overseas Technical Co-operation Agency, Tokyo, Japan.

Gupta, B.K.; Chatterjee, K.K. and Sinha, A.K. (1992) Floating behaviour of silkworm eggs. *Indian Silk*, 30: 24-26.

Gupta, B.K.; Kharoo, V.K. and Sahani, N.K. (1990) Effect of different types of laying sheets on number of eggs laid by silkworm (*Bombyx mori*). *Bioved*, 1: 79-80.

Gupta, B.K.; Sinha, A.K. and Das, B.C. (1991) Studies on egg yielding capacity of different races of silkworm (*Bombyx mori* L). *Giobios*, 18: 173-176.

Han, M.S. and Watanabe, H. (1987) Immunoperoxidase-staining method for discrimination of microsporidian spores in the pebrine infection of silkworm mother moths. *J. Seric. Sci. Japan*, 56: 431-435.

Han, M.S. and Watanabe, H. (1988) Transovarian transmission of two microsporidia in the silkworm, *Bombyx mori* and disease occurrence in the progeny population. *J. Invertebr. Pathol.*, 51: 41-45.

Han, M.S.; Nguyen, M.T. and Lim, J.S. (1997) Establishment of simplistic moth inspection system to prevent *Nosema bombycis* infection of the silkworm, *Bombyx mori*. *Korean J. Seric. Sci.*, 36: 69-75.

Haniffa, M.A. and Thatheyus, A.J. (1992) Effect of age related mating schedule on reproduction in the silkworm, *Bombyx mori* L. *Sericologia*, 32: 217-222.

Hartwig, A. and Mieczkowski, K. (1990) Diseases of silkworm – pebrine (*Nosema bombycis*) mycotic diseases. *Med. Weter.*, 46: 21-23.

Harvey, W.R. (1962) Metabolic aspects of insect diapause. *Ann. Rev. Ent.*, 7: 57-80.

Hasegawa, K. (1957) The diapause hormone of the silkworm, *Bombyx mori*. *Nature*, 179: 1300-1301.

Hatakeyama, Y. and Hayasaka, S. (2001) Specific amplification of microsporidia DNA fragments using multiprimer PCR. *J. Seric. Sci. Japan*, 70: 163-166.

Hayasaka, S. (1990) Inhibitory effect of high temperature on the development of *Nosema bombycis* in the silkworm, *Bombyx mori*. *Acta Seri. Entomol.*, 3: 59-65.

Hayasaka, S. and Ayuzawa, C. (1987) Diagnosis of microsporidians, *Nosema bombycis* and closely related species by antibody-sensitized latex. *J. Seric. Sci. Japan*, 56: 169-170.

Heinrich, B. (1981) A brief historical survey; in *Insect Thermoregulation*. Heinrich, B. (ed.), pp. 7-17, John Wiley, New York.

Hinton, H.E. (1969) Respiratory system of insect eggshells. *Ann. Rev. Ent.*, 14: 343-368.

Hinton, H.E. (1981) *Biology of Insect Eggs*. Vol. I, pp.413, Pregamon Press, Oxford.

Hitoshi, I. and Ishihara, R. (1981) Inhibitory effect of several chemicals against the hatch of *Nosema bombycis* spores. *J. Seric. Sci. Japan*, 50: 276-281.

Hodkova, M. and Hodek, I. (2004) Photoperiod, diapause and cold-hardiness. *European J. Entomol.*, 101: 445-458.

Honek, A. and Kocourek, F. (1990) Temperature and development time in insects: a general relationship between thermal constants. *Zoologische Jahrbilcher Systematik*, 117: 401-439.

Hurkadli, H.K. and Manjula, A. (1991) Artificial hatching of bivoltine silkworm eggs, *Bombyx mori* at different hours of oviposition for tropical conditions. (Lepidoptera: Bombycidae). *Sericologia*, 31: 345-347.

Hurkadli, H.K. and Manjula, A. (1992) Safe period of acid treatment for bivoltine silkworm (*Bombyx mori*) hybrid eggs at different hours of oviposition for varying temperatures. *Indian J. Seric.*, 31: 97-100.

Ikeda, M.; Su, Z.; Saito, H.; Imai, K.; Sato, Y.; Isobe, M. and Yamashita, O. (1993) Induction of embryonic diapause and stimulation of ovary trehalase activity in the silkworm, *Bombyx mori*, by synthetic diapause hormone. *J. Insect Physiol.*, 39: 889-895.

Indrasith, L.S.; Sasaki, T. and Yamashita, O. (1988a) A unique protease responsible for selective degradation of yolk protein in *Bombyx mori*–Purification, characterization and cleavage profiles. *J. Biol. Chem.*, 263:1045-1051.

Indrasith, L.S.; Izuhara, M.; Kobayashi, M. and Yamashita, O. (1988b) In-vitro translation of the protease catalising *Bombyx mori* egg specific protein and identification of peptide with biological activity. *Arch. Biochem. Biophy.*, 267: 328:333.

Irie, K. and Yamashita, O. (1980) Changes in vitellin and other yolk proteins during embryonic development in the silkworm, *Bombyx mori*. *J. Insect Physiol.*, 26: 811–817.

Ishihara, R. (1963) Effect of injection of *Nosema bombycis* (Nageli) on pupal development of the silkworm, *Bombyx mori* L. *J. Insect Pathol.*, 5: 131-140.

Ishihara, R. and Iwano, H. (1991) The lawn grass cutworm, *Spodoptera depravata* (Butter) as a natural reservoir of *Nosema bombycis* Nageli. *J. Seric. Sci. Japan*, 60: 236-337.

Isobe, M. and Goto, T. (1980) Diapause hormones; in *Neurohormonal Techniques in Insects*. Miller, T.A. (ed.), pp. 217-243, Springler Verlag, Berlin.

Iwano, H. and Ishihara, R. (1981) Inhibitions effect of several chemicals against hatching of *Nosema bombycis* spores. *J. Seric. Sci. Japan*, 50: 276-281.

Izumi, Y.; Sonoda, S.; Yoshida, H.; Danks, H.V. and Tsumuki, H. (2006) Role of membrane transport of water and glycerol in the freeze tolerance of the rice stem borer, *Chilo suppressalis* Walker (Lepidoptera: Pyralidae). *J. Insect Physiol.*, 52: 215-220.

Jadav, L.D. and Gajare, B.P. (1978) Studies on the effect of mating durations on the viability of silkworm (*Bombyx mori* L.) eggs. *Indian J. Seric.*, 17: 28-32.

Jaike, B. (1989) Investigation on the pigmented and non-pigmented eggs of a local silkworm race Madagascar. *Canye Kexue*, 15: 184-188.

Jayaswal, J.; Giridhar, K.; Somi Reddy, J. and Jagdish Prabhu, H. (2008) *Mulberry Silkworm Seed Production*. Published by Central Silk Board, India, pp. 132.

Jayaswal, K.P. (2002) Diapause phenomenon in tropical multivoltine races of *Bombyx mori* L.–A review. *Bull. Ind. Acad. Sci.*, 6: 1-20.

Jayaswal, K.P.; Singh, T. and Subba Rao, G. (1991) Effect of female pupal weight on fecundity of mulberry silkworm *Bombyx mori*. *Indian J. Seric.*, 30: 141–143.

Jolly, M.S. (1986) *Pebrine and its Control*. Published by Central Silk Board, Bangalore, India.

Jolly, M.S.; Subba Rao, S. and Krishnaswamy, S. (1964a) Studies on the mating capacity of male of mulberry silkworm and the possibility of utilizing polygamy in sericulture. *Indian J. Seric.*, 1: 25-32.

Jolly, M.S.; Subba Rao, S. and Krishnaswamy, S. (1964b) Effect of plane of inclination on egg laying by *Bombyx mori* L. *Indian J. Expt. Biol.*, 2: 165-166.

Kagawa, T. (1980) The efficacy of formalin as disinfectant of *Nosema bombycis* spores. *J. Seric. Sci. Japan*, 49: 218-222.

Kageyama, T. (1976) Pathways of carbohydrate metabolism in the eggs of the silkworm, *Bombyx mori. Insect Biochem.*, 6: 507-511.

Kai, H. and Haga, Y. (1978) Further studies on 'esterase A' in *Bombyx mori* eggs in relation to diapause development. *J. Seric. Sci. Japan*, 47: 125-133.

Kai, H. and Hasegawa, K. (1973) An esterase in relation to yolk cell lysis at diapause termination in the silkworm, *Bombyx mori. J. Insect Physiol.*, 19:799-810.

Kai, H.; Kawai, T. and Kawai, Y. (1987) A time interval activation of 'esterase A' by cold relation to the termination of embryonic diapause in the silkworm, *Bombyx mori. Insect Biochem.*, 17: 367-372.

Kai, H. and Nishi, K. (1976) Diapause development in *Bombyx mori* eggs in relation to 'esterase A' activity. *J. Insect Physiol.*, 22: 1315-1320.

Kalthoff, K. (1971) Position of targets and period of competence for UV-induction of the malformation 'double abdomen' in the eggs of *Smittia* species (Diptera: Chironomidae). *Roux's Arch.*, 168: 63-84.

Kamble, C.K. (1997) Fertility in *Bombyx mori* L. *Indian Text. J.*, 108: 56-59.

Kanada, T.; Matsumura H. and Ohtsuki Y. (1974) Surface structure of silkworm (*Bombyx mori*) egg. I. Fine structure of micropylar apparatus. *J. Seric. Sci. Japan*, 43: 379–383.

Kang, P.O.; Ryu, K.S.; Hong, K.W. and Sohn, B.H. (1993) Studies on occasional artificial hatching of hibernated eggs, *Bombyx mori* L. (I). The early hatching of post-hibernation of hibernated eggs, PDA. *J. Agric. Sci.* (Korea), 35: 725-729.

Kashiviswanathan, K. (1976) Sericulture industry in Japan. VI. Silkworm Breeding & Genetics. *Indian Silk*, 14: 13-21.

Katagiri, N.; Ando, O. and Yamashita, O. (1998) Reduction of glycogen in eggs of the silkworm, *Bombyx mori*, by use of a trehalase inhibitor, trehazalin and diapause induction in glycogen-reduced eggs. *J. Insect Physiol.*, 44: 1205-1212.

Kawaguchi, E. (1926) Polyspermy in *Bombyx mori* L. *Sangyo Simpo.*, 34: 38–41.

Kawamura, N. (1978) The early embryonic mitosis in normal and cooled eggs of the silkworm, *Bombyx mori*. *J. Morphol.*, 158: 57–72.

Kawamura, N. and Nakada, T. (1981) Studies on the increase in egg size in tetraploid silkworms induced from a normal and giant egg strain. *Jap. J. Genet.*, 56: 249-256.

Kawakami, Y.; Inoue, T.; Uchida, Y.; Hatakeyama, Y.; Iwano, H. and Ishihara, R. (1995) Specific amplification of DNA from different strains of *Nosema bombycis*. *J. Seric. Sci. Japan*, 64: 165-172.

Kawakami, Y.; Iwano, H.; Hatakeyama, Y.; Inoue, T.; Canning, E.U. and Ishihara, R. (2001) Use of PCR with specific primers for discrimination of *Nosema bombycis*. *J. Seric. Sci. Japan*, 70: 43-48.

Kawarabata, T. (2003) Review – Biology of microsporidians infecting the silkworm, *Bombyx mori* in Japan. *Insect Biochemistry & Sericology*, 72: 1-32.

Kawarabata, T. and Hayasaka, S. (1987) An enzyme-linked immunosorbent assay (ELISA) to detect alkali-soluble spore surface antigens of strains of *Nosema bombycis* (Microspora: Nosematidae). *J. Invertebr. Pathol.*, 50: 118-123.

Kawarabata, T. and Ishihara, R. (1984) Infection and development of *Nosema bombycis* in cell lines of *Antheraea eucalypti*. *J. Invertebr. Pathol.*, 44: 52-62.

Ke, Z.; Zie, W.; Wang, Z.; Long, Q. and Py, Z. (1990) A monoclonal antibody of *Nosema bombycis* and its use for identification of microsporidian species. *J. Invertebr. Pathol.*, 56: 395-400.

Kim, S.E.; Lee, S.M. and Seong, S.I. (1992) Long-term preservation of *Bombyx mori* stocks by frozen gonad storage. *Korean J. Seric. Sci.*, 34: 1-7.

Kim, S.E.; Shikata, M. and Akai, H. (1981) Eggshell lipids in relation to water evaporation and diapause in the silkworm, *Bombyx mori*. *J. Seric. Sci. Japan*, 50: 94–100.

Kishore, S.; Baig, M.; Nataraju, B.; Balavenkatasubbaiah, M.; Sivaprasad, D.; Iyengar, M.N.S. and Datta, R.K. (1994) Cross infectivity of microsporidians isolated from wild lepidopteron insects to silkworm, *Bombyx mori* L. *Indian J. Seric.*, 33: 126-130.

Kitazawa, T.; Kanada, T. and Takami, T. (1963) Changes in the mitotic activity in the silkworm eggs in relation to diapause. *Bull. Seric. Exp. Stn.*, 18: 283-295.

Kitazawa, T.; Kiguchi, K.; Sugiyama, H. and Ohtsuki, Y. (1989) Selection of egg sensitivity to the hydrochloric acid treatment and its effect to diapause in the silkworm, *Bombyx mori*. *J. Seric. Sci. Japan*, 58: 511-516.

Kittlans, E. (1961) Die Ebmryohalant wicklung von *Leptinotarsa decemlineata*, *Epilachna sparsa* and *Epilachna vigintiocto maculata* in abhangigkeit von der temperature. *Deutsche Ent.*, 8: 41-52.

Kovalev, P.A. (1970) *Silkworm Breeding Stocks*. Published by Central Silk Board, Bangalore, India.

Kramer, J.P. (1976) The extra corporal ecology of microsporidia; in *Comprehensive Pathology*. Bulla L.A. and Cheng, L.A. (eds.), Vol. I. Pp. 127-135, Plenum Press, Publishing Corporation, New York.

Krishnaswami, S.; Narasimhanna, M.N.; Suryanarayan, S.R. and Kumararaj, S. (1973) *Sericulture Manual-II – Silkworm Rearing*. FAO Agricultural Services Bulletin, 15/3, Rome, Italy.

Kubota, I.; Isobe, M.; Imai, K.; Goto, T.; Yamashita, O. and Hasegawa, K. (1979) Characterization of the silkworm diapause hormone-B. *Agric. Biol. Chem.*, 43: 1075-1078.

Kumar, K.C.M. and Paul, D.C. (1993) Effect of age of male on ejaculation and behaviour of spermatozoa in the internal reproductive organs of the female silk moth, *Bombyx mori* L. (Lepidoptera: Bombycidae). *Sericologia*, 33: 219-222.

Kunkel, J.G. and Nordin, J.H. (1985) Yolk Proteins; in *Comprehensive Insect Physiology, Biochemistry and Pharmacology*. Kerkut, G.A. and Gilbert, L.I. (eds.), Vol. I., pp. 83 -111, Pergamon Press, Oxford.

Kurata, S.; Koga, K. and Sakaguchi, B. (1979) RNA content and RNA synthesis in diapause and non-diapause eggs of *Bombyx mori*. *Insect Biochem.*, 9: 107-110.

Kurata, S.; Yaginuma, T.; Kobayashi, Y.; Koga, K. and Sakaguchi, B. (1980) DNA content and cell number during the embryogenesis of *Bombyx mori*. *J. Seric. Sci. Japan*, 49: 107–110.

Kurisu, K. (1986) Simplified calculation of the operating characteristics in the pebrine inspection of the grouping mother method. *J. Seric. Sci. Japan*, 55: 351-352.

Kurisu, K.; Nakasone, S.; Dohi, M. and Hamazaki, M. (1985) Studies on the pebrine inspection of the mother moth in the silkworm, *Bombyx mori* for commercial eggs. III. Practical application of sampling inspection table. *Bull. Fac. Text. Sci. Kyoto Univ.*, 11: 19-30.

Kuroida, K. and Nihei, K. (1979) Studies on the cold storage of silkworm eggs treated with hydrochloric acid after chilling. *J. Seric. Sci. Japan*, 48: 401-403.

Laudien, H. (1973) Changing reaction systems; in *Temperature and Life*. Precht, H.; Christopher, J.; Hensel, H. and Larcher, W. (eds.), pp 355-399, Springer Verlag Berlin.

Lee, R.E. Jr. (1991) Principles of insect low temperature tolerance; in *Insects at Low Temperature*. Lee, R.E. Jr. and Denlinger, D.L. (eds.), pp 17-46, Chapman and Hall, New York.

Legay, J.M. ('1989) Contribution to the typological analysis of the laying behavior in *Bombyx mori* L. *Sericologia*, 29: 163-166.

Leopold, R.A. (1976) The role of male accessory glands in insect reproduction. *Annu. Rev. Entomol*, 21: 199-221.

Li, D. (1985) Studies on the serological diagnosis of the pebrine of silkworm, *Bombyx mori*, by slide agglutination test. *Sci. Seric.*, 11: 99 -102.

Linzen, B. (1974) The tryptophane–ommochrome pathway in insects. *Adv. Insect Physiol.*, 10:117-246.

Liu, S.X. (1984) A summary of the techniques for the control of *Nosema bombycis* in *Bombyx mori* with instant acid treatment at high temperature. *Guangdong Agric. Sci.*, 3: 20-21.

Liu, S.X. and Zhong, W.B. (1988) The research channels in the prevention and control of silkworm diseases. *Sericologia*, 29: 287-295.

Liu, S.X.; Zhu, D.Z.; Zheng, W.Y.; Huo, Y.C.; Wu, Y.N. and Liang, J.Y. (1971) A summary of the techniques for the control of *Nosema bombycis* Negeli in *Bombyx mori* with instant acid treatment at high temperature. *Guangdong Agric Sci.*, 3: 20-21.

Loasesthakul, W. and Kuribayashi, S. (1976) Sex separation of silkworm pupa by means of brine assortment. *Bull. Thai Seric. Res. & Training Center*, 6: 81-83.

Lytle, D.A. (2002) Flash floods and aquatic insect life-history evolution: evaluation of multiple models. *Ecology*, 83: 370-385.

Malone, L.A. and Mclvor, C.A. (1995) DNA probes for two microsporidia, *Nosema bombycis* and *Nosema costelytrae*. *J. Invertebr. Pathol.*, 65: 269-273.

Malone, L.A.; Broadwell, A.H.; Lindridge, E.T.; McIvor, C.A. and Ninham, J.A. (1994) Ribosomal RNA genes of two microsporidia, *Nosema apis* and *Vavraia oncoperae*, are very variable. *J. Invertebr. Pathol.*, 64:151-152.

Manjula, A. (1993) Oviposition behaviour in silkworm moths under tropical conditions. *Indian Text. J.*, 103: 102-104.

Manjula, A. and Hurkadli, H.K. (1986) Cold acid treatment of bivoltine silkworm eggs, *Bombyx mori* for tropical country. *Sericologia*, 26: 26-29.

Manjula, A. and Hurkadli, H.K. (1993) Effect of cold storage of multivoltine and multi x bivoltine silkworm eggs, *Bombyx mori* L. (Lepidoptera: Bombycidae) at low temperature on hatching performance. *Sericologia*, 33: 615-620.

Manjula, A. and Hurkadli, H.K. (1995) Chilling of silkworm eggs. *Indian Text. J.*, 106: 70-73.

Manjula, A. and Hurkadli, H.K. (1996) Studies on the hibernation schedule for bivoltine pure and hybrid silkworm eggs, *Bombyx mori* (Lepidoptera: Bombycidae) aestivated from twenty days under tropical conditions. *Sericologia*, 36: 665-675.

Manjulakumari, D. and Geethabali (1991) Ovipositional response of *Bombyx mori* (L) to different textures of the substratum. *Entomon*, 16: 11-15.

Margaritis, L.H. (1985) Structure and physiology of eggshell; in *Comprehensive Insect Physiology, Biochemistry and Pharmacology.*

Kerkut, G.A. and Gilbert, L.I. (eds.), Vol. I. pp.153 -230, Pergamon Press, Oxford.

Mathur, S.K.; Kumar, K.; Rao, G.S.; Singh, B.D. and Lall, S.B. (1992) Hot water treatment to break diapause in silkworm eggs. *Indian Text. J.*, 103: 46-48.

Mathur, S.K. and Lal, S.B. (1993) Effects of temperature and humidity on the adaptability of insects. *Indian Textile J.*, 136: 34-47.

Mathur, S.K.; Roy, A.K. and Mukherjee, P.K. (1988) Studies on the effect of number of eggs per laying on the development of silkworm race -Nistari, *Bombyx mori* L. (Lepidoptera: Bombycidae). *Indian J. Seric.*, 27: 168-170.

Mathur, S.K.; Roy, A.K.; Sen, S.K. and Subba Rao, G. (1995) Fecundity and *Bombyx mori* L.–Light effects on silkworms. *Indian Text. J.*, 106: 74-76.

Mathur, S.K. and Singh, T. (1990) Abnormal eclosion in silkworm, *Bombyx mori*. *Indian Text. J.*, 100: 150-151.

Mathur, V.B.; Singh, G.P. and Kamble, C.K. (1995) Effect of photoperiod on the rearing and diapause in the silkworm, *Bombyx mori* L. *J. Seric.*, 3: 46-49.

Matsuzaki, M. (1968) Electron microscopic studies on the chorion formation of the silkworm, *Bombyx mori*. *J. Seric. Sci. Japan*, 37: 483 -490.

Mazur, G.D.; Regier, J.C. and Kafatos, F.C. (1982) Order and defects in the silk moth chorion, a biological analogue of a cholesteric liquid crystal; in *Insect Ultrastructure.* King, R.C. and Akai, H. (eds.), Plenum Press, New York, 1: 150-185.

Mehrunnessa, J. and Barman, A.C. (1996) Simplified method of hot water treatment for breaking diapause of bivoltine silkworm eggs. *Bull. Seric. Res.*, 7: 91-93.

Mei, L. and Jin, W. (1998) Study on distinguishing *Nosema bombycis* by SPA co-agglutination. *Sci. Seric.*, 14: 110-111.

Mill, P.J. (1985) Structure and physiology of respiratory system; in *Comprehensive Insect Physiology, Biochemistry and Pharmacology.* Kerkut, G.A. and Gilbert, L.I. (eds.), Vol. III., pp. 518–593, Pergamon Press, Oxford.

Ming, W.S. (1994) *Silkworm Egg Production*. Vol. III., Food and Agriculture Organization, Agricultural Services Bulletin, United Nations, Oxford & IBH Publishing Co. Pvt. Ltd., New Delhi, India.

Miya, K. (1978) Electron microscopic studies on the early embryonic development of the silkworm, *Bombyx mori*. I. Architecture of the newly laid eggs and the changes by sperm entry. *J. Fac. Agric. Iwate Univ. Japan*, 14: 11–35.

Miya, K. (1984) Early embryogenesis of *Bombyx mori;* in *Insect Ultrastructure*. King, R.C. and Akai, H. (eds.), Vol. II., pp. 49 -72, Plenum Press, New York.

Miya, K.; Kurihara M. and Tanimura I. (1972) Changes in the fine structure of the serosa cell and the yolk cell during diapause and post-diapause development in the silkworm, *Bombyx mori* L. *J. Fac. Agric. Iwate Univ. Japan*, 8:11–87.

Mizuno, T. (1920) On the cold storage of silkworm eggs. *Sanshi Shikenjo Hokoku*, 4: 205-262.

Mizuta, Y.; Honouse, T. and Wada, T. (1969) Relationship between weight of pupae and number of eggs by moth of silkworm, *Bombyx mori*. *Bull. Seric. Expt. Stn.*, 23: 515-532.

Morohoshi, S. and Mezaki, M. (1957) Studies on the voltinism in the domestic silkworm. IV. On the locus of the genes controlling voltinism. *J. Seric. Sci. Japan*, 26: 257.

Morohoshi, T. (1979) *Developmental Physiology of the Silkworm*. Gakkai Shupan Centre, Tokyo.

Müller, A.; Trammer, T.; Chioralia, G.; Seitz, H.M.; Diehl, V. and Franzen, C. (2000) Ribosomal RNA of *Nosema algerae* and phylogenetic relationship to other microsporidia. *Parasitol. Res.*, 86: 18-23.

Nagatomo (1953) On the significance of Im-type in the silkworm. *Japanese J. Breed.*, 13: 60.

Nageli, K.W. (1857) Uber die neue Krankheit der Seidenraupe und verwandte Organismen. *Bot. Z.*, 15: 760-761.

Nageswararao, S.; Muthulakshmi, M.; Kanginakudru, S. and Nagaraju, J. (2004) Phylogenetic relationships of three new microsporidians isolated from the silkworm, *Bombyx mori*. *J. Invertebr. Pathol.*, 86: 87-95.

Nageswararao, S.; Surendranath, B. and Saratchandra, B. (2005) Characterization and phylogenetic relationships among microsporidia infecting silkworm, *Bombyx mori,* using inter simple sequence repeat (ISSR) and small subunit rRNA (SSU-rRNA) sequence analysis. *Genome,* 48: 355-366.

Nakada, T. (1932) The resistance of silkworm embryo to abnormal high temperature and dryness on the external morphology of the embryo. *Bull. Fukuoka Sweicult. Exp. Stn.,* 1: 1 -26.

Nakamura, K. and Numata, H. (2000) Photoperiodic control of the intensity of diapause and diapause development in the bean bug, *Riptortus clavatus* (Heteroptera: Alydidae). *European J. Entomol.,* 97: 19-23.

Nakasone, S. (1979) Changes in lipid components during the embryonic development of the silkworm, *Bombyx mori. J. Seric. Sci. Japan,* 48: 526–532.

Narasimhanna, M.N. (1988) *Manual on Silkworm Eggs Production.* Published by Central Silk Board, Bangalore, India, pp. 169.

Narayanaswamy, T.K.; Surendra, H.S. and Govindan, R. (2000) Relationship between weight components and egg parameters in silkworm, *Bombyx mori. Proc. Natl. Semi. Trop. Seric.,* Vol. II., pp. 79-81.

Nataraju, B.; Sathyaprasad, K.; Manjunath, D. and Aswani Kumar, C. (2005) *Silkworm Crop Protection.* Published by Central Silk Board, Bangalore, India, pp. 412.

Obara, Y. (1979) *Bombyx mori* mating dance: An essential in locating the females. *Appl. Entomol. Zool.,* 14: 130-132.

Ohio, H.; Miyahara, J. and Yamashita, A. (1970) Analysis of practically important characteristics in the silkworm in early breeding generations of hybrids. Variations among strains, correlation between parents and offspring as well as relation between each character. *Tech. Bull. Seric. Expt. Stn. Japan,* 93: 39-49.

Ohtsu, T.; Kimura, M.T. and Katagiri, C. (1998) How *Drosophilla* species acquire cold tolerance: qualitative changes of phospholipids. *European J. Biochem.,* 252: 608-611.

Ohtsuki, R.; Sato, S.; Nagai, I.; Horigone, N.; Yoshitake, S.; Goto, s.; Totani, C.; Nagashima, E. and Yoshitake, S. (1997) *Silkworm*

Egg Production (translated from Japanese), Oxford & IBH Publishing Co. Pvt. Ltd., New Delhi.

Ohtsuki, Y.; Kanada, T. and Matsumura, H. (1977) Surface structure of silkworm (*Bombyx mori*) eggs. II. Characteristics of different areas of egg surface. *J. Seric. Sci. Japan*, 46: 45–50.

Ohtsuki, Y. and Murakami, A. (1968) Nuclear division in the early embryonic development of the silkworm, *Bombyx mori* L. *Zool. Mag.*, 77: 383–387.

Ohtsuki, Y.; Mori, S.; Kanada, T. and Kitazawa, T. (1976) Morphological observation on the embryonic moult in the silkworm, *Bombyx mori. J. Seric. Sci. Japan*, 45: 225–231.

Okada, M. (1970a) Electron microscopic studies on diapause embryos of the silkworm, *Bombyx mori* L. *Sci. Rep. Tokyo Kvoiku Daigdku* Sect B, 14: 95–111.

Okada, M. (1970b) Changes in affinity of lysosomes to acridine organge during diapause and development of embryos of the silkworm, *Bombyx mori* L *Zool. Mag.*, 79: 204–210.

Okada, M. (1971) Role of chorion as a barrier to oxygen in the diapause of the silkworm, *Bombyx mori* L. *Experientia*, 27: 658-660.

Omura, S. (1939) Oviposition mechanism of the silkworm moths. I. Stimulation to elicit the oviposition. *J. Seric. Sci. Japan*, 10: 47-49.

Ono, T. (1980) Role of the scales as a releaser of the copulation attempt in the silkworm moth (*Bombyx mori*). *Kontyu*, 48: 540-544.

Osanai, M. and Yonezawa, Y. (1986) Changes in amino acid pool in the silkworm, *Bombyx mori* during embryonic life. Alanine accumulation and its conversion to proline during diapause. *Insect Biochem.*, 16: 373-379.

Osanai, M.; Kasuga, H. and Aigaki, T. (1987) Physiological role of apyrene spermatozoa of *Bombyx mori. Experientia*, 43: 593-596.

Osanai, M.; Kasuga, H. and Aigaki, T. (1989) Isolation of eupyrene sperm bundles and apyrene spermatozoa from seminal fluid of the silk moth, *Bombyx mori. J. Insect Physiol.*, 35: 401-408.

Osanai, M.; Kasuga, H. and Aigaki, T. (1990) Physiology of sperm maturation in the spermatophore of the silkworm, *Bombyx mori. Advances Invertebr. Reprod.*, 5: 531-536.

Ovanesyan, T.T. and Lobzhanidze, V.I. (1960) First results of experiments on thermal disinfection on pebrinous silkworm eggs by a brief immersion in hot water. *Inst Mortl Zhivotnykh Akad Nauk SSSR*, 21: 184-215.

Pallavi, S.N. and Kamble, C.K. (1997) Disinfections and hygiene in sericulture–A review. *Sericologia*, 37: 401-415.

Pasteur, L. (1870) Etudes sur la maladie des vers a soie. P. 322, Gauthier-Villars, Imprimeur-Libraire, Paris.

Patil, C.M. and Gowda, B.L.V. (1986) Environmental adjustment in sericulture. *Indian Silk*, 25: 11-14.

Patil, C.S. (1993) Review on pebrine–a microsporidian disease in the silkworm, *Bombyx mori* L. *Sericologia*, 33: 201-210.

Patil, C.S. and Geethabai, M. (1989) Studies on the susceptibility of silkworm races to pebrine spores. *J. Appl. Entomol.*, 108: 421-423.

Patil, C.S. and Geethabai, M. (1985) *Guidelines for identification of pebrine spores in grainages.* Published by KSSRDI, Bangalore, India.

Patil, C.S. and Geethabai, M. (1989) Studies on the susceptibility of silkworm races to pebrine spores. *J. Appl. Entomol.*, 108: 421-423.

Patil, C.S.; Jyothi, N. B. and Dass, C.M.S. (2001) Silkworm faecal pellets examination as diagnostic method for detecting pebrine. *Indian Silk*, 39: 11-12.

Paul, D.C. and Kishor Kumar, C.M. (1995) Influence of male age on mating capacity, fecundity and fertility of mated female silk moth, *Bombyx mori* L. under high temperature and high humidity conditions. *Entomon*, 20: 253-255.

Paul, D.C.; Kishor Kumar, C.M. and Sen, S.K. (1993) Effect of age at mating on oviposition, fertility and longevity of female silk moth, *Bombyx mori* L. (Lepidoptera: Bombycidae). *Indian J. Seric.*, 32: 24-25.

Perk, K.E. (1973) The fine structure of the diapause regulator cell in the suboesophageal ganglion in the silkworm, *Bombyx mori*. *J. Insect Physiol.*, 19: 293-302.

Perk K.E. and Yoshitake N. (1970a) A radio autographic study of diapause in the silkworm, *Bombyx mori*. *J. Insect Physiol.*, 16: 1655-1663.

Perk, K.E. and Yoshitake N. (1970b) Function of the embryo and yolk cells in diapause of the silkworm eggs (*Bombyx mori*). *J. Insect Physiol.*, 16: 2223-2239.

Peters, T.M. and Barbosa, P. (1977) Influence of population density on size, fecundity and developmental rate of insect in culture. *Rev. Entomol.*, 22: 431-450.

Peter, A.; Sadatulla, F. and Devaiah, M.C. (1999) The viral, bacterial and protozoan diseases of the silkworm, *Bombyx mori* L.; in *Advances in Mulberry Sericulture*. Devaiah, M.C.; Narayanaswamy, K.C. and Maribashetty, V.G. (eds.), pp. 378-457, C.V.G. Publications, Bangalore, India.

Petkov, N. and Mladenov, G. (1979) Studies on the multiple use of silk moth (*Bombyx mori*) male individuals in the production of pedigree and commercial silkworm seed. *Anim. Sci.*, 16: 107-115.

Petkov, N.; Yolov, A.; Mladenov, G. and Nacheva, I. (1979) Influence of mating length of silk moths of some inbred silkworm (*Bombyx mori*) line on silkworm seed quality and quantity. *Anim. Sci.*, 16: 116-122.

Pitts, K.M. and Wall, R. (2006) Cold shock and cold tolerance in larvae and pupae of the blow fly, *Lucilia sericata*. *Physiological Entomol.*, 31: 57-62.

Powell, G.; Tosh, C.R. and Hardie, J. (2006) Host plant selection by aphids: behavioral, evolutionary and applied perspectives. *Ann. Rev. Entomol.*, 51: 309-330.

Prowse, T.D. and Culp, J.M. (2003) Ice breakup: a neglected factor in river ecology. *Can. J. Civic. Engineering*, 30: 128-144.

Punitham, M.T.; Haniffa, M.A. and Arunachalam, S. (1987) Effect of mating duration on fecundity and fertility of eggs in *Bombyx mori* (Lepidoptera: Bombycidae). *Entomon*, 12: 5-8.

Quadrefague, A De (1860) cit. in; *Pebrine diseases of silkworms*–a technical report, Tatsuke K (1971) Overseas Technical Co-operation Agency, Tokyo, Japan.

Rahman, S.M. and Ahmad, S.U. (1989) Artificial hatching of hibernated eggs of the silkworm, *Bombyx mori* L. by warm water treatment. *Bangladesh J. Zool.*, 17: 117-122.

Rahman, S.M.; Raza, M.A.S.; Salem, M.A. and Bari, A. (1991) Effect of larval population density on the fecundity and hatchability of silkworm, *Bombyx mori* (L). *Bull. Seric. Res.*, 2: 7-12.

Raju, C.S.; Thiagarajan, V.; Vemananda Reddy, G. and Suryanarayana, N. (1990) Studies on the post-mating refrigeration of the female moth of the silkworm, *Bombyx mori* L. *Indian J. Seric.*, 29, 213-218.

Raju, S.; Dharmapandit, S.K. and Datta, R.K. (1993) A simple technique of sex separation in mulberry silkworm. *Indian Silk*, 31:16-18.

Ram, K. and Singh, D. (1992) Role of mating disruption in the production of viable silkworm (*Bombyx mori*) eggs. *J. Entomol. Res.*, 16: 206-210.

Ramlov, H. (2000) Aspects of natural cold tolerance in ectothermic animals. *Human Reproduction*, 15: 26-46.

Ravindra S.; Chaturvedi, H.K. and Datta, R.K. (1994) Fecundity of mulberry silkworm, *Bombyx mori* L. in relation to female cocoon weight and repeated mating. *Indian J. Seric.*, 33: 70-71.

Ravindra S.; Nagaraju, J. and Vijayaraghavan, K. (1990) Effect of hot water on the prevention of diapause in the silkworm, *Bombyx mori*. *Sericologia*, 30: 443-447.

Reddy, V.; Rao, V. and Kamble, C.K. (2005) *Manual on Silkworm Eggs, Bombyx mori*. Published by Silkworm Seed Technology Laboratory, Central Silk Board, Bangalore, India.

Regier, J.C. and Kafatos, F.C. (1985) Molecular aspects of chorion formation; in *Comprehensive Insect Physiology, Biochemistry and Pharmacology*. Kerkut, G.A. and Gilbert, L.I. (eds.), Vol. I., pp. 113–151, Pergamon press, Oxford, New York.

Regier, J.C.; Mazur G.D. and Kafatos, F.C. (1980) The silk moth chorion: Morphological and biochemical characterization of four surface regions. *Devel. Biol.*, 76: 286–304.

Relina, L.I. and Gulevsky, A.K. (2003) A possible role of molecular chaperones in cold adaptation. *CryoLetters*, 24: 203-212.

Renault, D.; Nedved, O.; Hervant, F. and Vernon, P. (2004) The importance of fluctuating thermal regimes for repairing chill injuries in the tropical beetle *Alphitobius diaperinus* (Coleoptera: Tenebrionidae) during exposure to low temperature. *Physiological Entomol.*, 29: 139-145.

Retnakaran, A. and Percy, J. (1985) Fertilization and special mode of reproduction; in *Comprehensive Insect Physiology, Biochemistry and Pharmacology*. Kerkut, G.A. and Gilbert, L.I. (eds.), Vol. I, pp. 242-253, Pergamon press, Oxford, New York.

Ring, R.A. and Danks, H.V. (1994) Desiccation and cryoprotection: overlapping adaptations. *CryoLetters*, 19: 275-282.

Sakaguchi, B. (1978) Gametogenesis, fertilization and embryogenesis of silkworm; in *The Silkworm: An Important Laboratory Tool*. Tazima, Y. (ed.), pp. 132-162.

Salt, R.W. (1959) Role of glycerol in the cold-hardiness of *Bracon cephi*. *Can. J. Zool.*, 37: 59-69.

Salt, R.W. (1961) Principles of insect cold-hardiness. *Ann. Rev. Ent.*, 6: 55-74.

Samson, M.V. (2000) Cocoon production and silkworm protection. *National Conference on Strategies for Sericulture Research & Development*, 16-18 November, CSR&TI Mysore, India, pp.38-48.

Samson, M.V. and Biram Saheb, N.M. (1999) Silkworm egg: key to silk industry; in *Advances in Mulberry Sericulture*. Devaiah, M.C.; Narayanaswamy, K.C. and Maribashetty, V.G. (eds.), pp. 243-292, CVG Publications, Bangalore, India.

Samson, M.V.; Santha, P.C.; Singh, R.N. and Sasidharan, T.O. (1999a) A new microsporidian infecting *Bombyx mori* L. *Indian Silk*, 37: 10-12.

Samson, M.V.; Santha, P.C.; Singh, R.N. and Sasidharan, T.O. (1999b) Microsporidian spore isolated from *Pieris* sp. *Indian Silk*, 38: 5-8

Sander, K.; Guteit, H.O. and Jackle, H. (1985) Insect embryogenesis: morphology, physiology, genetical and molecular aspects; in *Comparative Insect Physiology, Biochemistry and Pharmacology*. Kerkut, G.A. and Gilbert, L.I. (eds.), Vol. I, pp. 321-385, Pergamon Press, Oxford.

Santha, P.C.; Sasidharan, T.O.; Singh, R.N.; Daniel, A.G.K. and Veeraiah, T.M. (2001) Identification of intermediary stages of *Nosema bombycis* for diagnosis of pebrine–a new approach. *Indian Silk*, 40: 13-14.

Sasidharan, T.O.; Singh, R.N.; Samson, M.V.; Manjula, A.; Santha, P.C. and Chandrashekharaiah, A. (1994) Spore replication rate of *Nosema bombycis* (Microsporidia: Nosematidae) in the silkworm, *Bombyx mori* L. in relation to pupal development and age of moths. *Insect Sci. Applic.*, 15: 427-431.

Sato, R. and Watanabe, H. (1980) Purification of mature microsporidian spores by isodensity equilibrium centrifugation. *J. Seric. Sci. Japan*, 49: 512-516.

Sato, R. and Watanabe, H. (1986) Pathways of oral infection with four microsporideae in the silkworm, *Bombyx mori* L. *J. Seric. Sci. Japan*, 55:10-26.

Sato, R.; Kobayashi, M. and Watanabe, H. (1981) Internal ultra structure of spores of microsporidians isolated from the silkworm, *Bombyx mori* L. *J. Invertebr. Pathol.*, 40: 260-265.

Sato, R.; Masahiko, K.; Watanabe, H. and Fujiwara, T. (1982) Serological discrimination of several kinds of microsporidian spores isolated from the silkworm, *Bombyx mori* by an indirect fluorescent antibody technique. *J. Seric. Sci. Japan*, 50: 180-184.

Scriber, J.M. and Slarisky, F. Jr. (1981) The nutritional ecology of immature insects. *Ann. Rev. Entomol.*, 26: 181-211.

Sengupta, K.; Datta, R.K. and Biswas, S.N. (1973) Effect of multiple crossing on the type of progeny recovered in silkworm, *Bombyx mori* L. *Indian J. Seric.*, 12: 31-38.

Sengupta, K.; Kumar, P.; Baig, M. and Govindaiah (1990) *Handbook of Diseases and Pests of Silkworm and Mulberry*. UNESCAP, Bangkok, Thailand.

Shaheen, A.; Trag, A.R.; Nabi, G. and Ahmad, F. (1992) Correlation between female pupal weight and fecundity in bivoltine silkworm, *Bombyx mori* L. *Entomon*, 17: 109-111.

Shapiro, B.M.; Schackmann, R.W. and Gabel C.A (1981) Molecular approaches to the study of fertilization. *Annu. Rev. Biochem.*, 50:815-843.

Sheeba, R.; Devaiah, M.C.; Chinaswamy, K.P. and Govindan, R. (1999) Thermotherapy of pebrinized cocoons and its effect on larval progeny; in *Proc. Natl. Semi. Trop. Seri.* Govindan, R.; Chinaswamy, K.P.; Krishnaprasad, N.K. and Reddy, D.N.R. (eds), Vol. II., pp. 266-272, UAS and Swiss Agency for Development and Cooperation, Bangalore, India.

Shenyuan, P.; Liguo, G.; Lixia, L; Xiaohui, T.; Simei, H. and Ming, T. (1998) An approach to the method of automatic sex discrimination by single cocoon scaling. *Acta Sericologica Sinica*, 24: 91-94.

Shi, L. and Jin, P. (1997) Study on the differential diagnosis of *Nosema bombycis* of the silkworm, *Bombyx mori* by monoclonal antibody-sensitized latex. *Sericologia*, 37: 1-6.

Shiqing, X.; Guoqin, T.; Hidennori, K.; Junliang, X.; Biping, Z.; Xilin, C. and Yanghu, S. (2000) Change of TAPase activity for esterase-A *in-vitro* 25°C and its relationship with diapause in silkworm eggs, *Bombyx mori. Canye Kexue*, 26: 140-145.

Siddhu, N.S.; Sreenivasan, R. and Shamachary, A. (1967) Fertility performance of female moths depends on their male moths. *Indian J. Seric.*, 3: 77-84.

Singh, G. P. and Kumar, V. (1995) Correlation of larval and cocoon weight with fecundity in silkworm, *Bombyx mori* L. *Ann. Entomol.*, 13: 15-17.

Singh, G.P.; Mathur, V.B.; Kamble, C.K. and Datta, R.K. (1994) Cold storage of bivoltine female silkworm moths. *Indian Silk*, 33: 25-26.

Singh, G.P.; Mathur, V.B.; Vineet Kumar; Kamble, C.K. and Datta, R.K. (1994) Refrigeration of virgin female moths of silkworm, *Bombyx mori* L. (Lepidoptera: Bombycidae) and its impact on fecundity, viability and hatchability. *Uttar Pradesh J. Zool.*, 14: 133-136.

Singh, G. P.; Vineet Kumar; Mathur, V.B. and Kamble, C.K. (1995) Synergistic action of acid treatment and surface sterilization. *Indian Silk*, 34: 25-25.

Singh, K. and Shukla, G.S. (1992) Ready reckoner for hydrochloric acid treatment of silkworm eggs of weak diapause character. *Sericologia*, 32: 51-56.

Singh, R.N.; Daniel, A.G.K.; Sindaggi, S.S. and Kamble, C.K. (2007) Microsporidians infecting silkworm, *Bombyx mori*. *Sericologia*, 47: 1-16.

Singh, R.N.; Yadav, P.R. and Singh, T. (1992) Containing the Pebrine problem in sericulture. *Indian Silk*, 30: 43–44.

Singh, T. (1989) Diapause rather a complicated phenomenon. *Indian Silk*, 28: 2-3.

Singh, T. (1998) Behavioural aspects of oviposition in silkworm *Bombyx mori*. *Indian J. Seric.*, 37: 101-108.

Singh, T. (1994) Correlation between pupal weight and fecundity in *Bombyx mori* (L). *Annl. Ent.*, 12: 5 – 7.

Singh, T. and Mathur, S.K. (1989) The dancing of male silk moths to the perfume to female moth. *Indian Silk*, 28: 17-18.

Singh, T. and Rao, G.S. (1994) Studies on the evaluation of hybrids by breeding index in *Bombyx mori* L. *Entomon*, 19: 169-170.

Singh, T. and Samson, M.V. (1999) Embryonic diapause and metabolic changes during embryogenesis in mulberry silkworm, *Bombyx mori* L. *J. Seric.*, 7: 1-11.

Singh, T. and Saratchandra, B. (2002) Biochemical changes during embryonic diapause in mulberry silkworm, *Bombyx mori* L. (Lepidoptera: Bombycidae). *Intl. J. Indust. Entomol.* (Korea), 5: 1-12.

Singh, T. and Saratchandra, B. (2003) Microsporidian disease of the silkworm, *Bombyx mori* L. (Lepidoptera: Bombycidae). *Intl. J. Indust. Entomol.* (Korea), 6: 1-9.

Singh, T. and Saratchandra, B. (2004) *Principles and Techniques of Silkworm Seed Production*. Discovery Publishing House, New Delhi, India, pp. 361.

Singh, T.; Bhat. M.M. and Khan, M.A. (2009) Adaptation in insects to changing environments–temperature and humidity. *Intl. J. Indust. Entomol.* (Korea), 19: 155-164.

Singh, T.; Saratchandra, B. and Phani Raj, H.S. (2003) Physiological and biochemical modulations during oviposition and egg laying in the silkworm, *Bombyx mori* (L). *Int. J. Indust. Entomol.*, 6: 115-123.

Singh, T. and Tripathi, P.M. (1995) Studies on mating duration and its effect on fecundity and fertility in mulberry silkworm moth *Bombyx mori*. (Lepidoptera: Bombycidae). *Bioved*, 6: 15-18

Singh, T.; Rao, V.; Jingade, A.H. and Samson, M.V. (2002) Structure and development of silkworm (*Bombyx mori*) eggs. *Sericologia*, 43: 323-334.

Sironmani, A. (1997) Detection of *Nosema bombycis* infection in the silkworm, *Bombyx mori* by Western blot analysis. *Sericologia*, 39: 209-216.

Slachta, M.; Vambera, J.; Zahradnckova, H. and Kostal, V. (2002) Entering diapause is a prerequisite for successful cold-acclimation in adult, *Graphosoma lineatum* (Heteroptera: Pentatomidae). *J. Insect Physiol.*, 48: 1031-1039.

Smyk, D. (1959) Methodes physiques de lutte contre *Nosema bombycis* dans les graines du ver a soie du murier. *Revue du Ver a Soie.*, 11: 155 161.

Somme, L. (1982) Super-cooling and winter survival in terrestrial arthropods. *Comp. Biochem. Physiol.*, 73(A): 519-543.

Sonobe, H.; Maotaini, K. and Nakajima, H. (1986) Studies on the embryonic diapause in the PND mutant of the silkworm, *Bombyx mori*. Genetic control of embryogenesis. *J. Insect Physiol.*, 32: 215-220.

Sonobe, H.; Matsumoto, A.; Fukuzaki, Y. and Fujiwara, S. (1979) Carbohydrate metabolism and restricted oxygen supply in the eggs of the silkworm, *Bombyx mori*: Genetic control of embryogenesis. *J. Insect Physiol.*, 25: 215-220.

Sprague, V. (1982) *Microspora;* in *Synopsis and Classification of Living Organism*. Parker, S.P. (ed.), Vol. 1, pp.589-594, McGraw Hill Book Co., New York.

Srikanta, H.K. (1986) Studies on the cross infectivity and viability of *Nosema bombycis* (Microsporidia: Nosematidae). M. Sc. Thesis UAS Bangalore, India, pp. 106.

Strunnikov, V.A. (1983) *Control of Silkworm Reproduction*, Development and Sex. MIR Publishers, Moscow, 280 pp.

Su, Z.H.; Ikeda, M.; Sato, Y.; Saito, H.; Imai, K.; Isobe, M. and Yamashita, O. (1994) Molecular characterization of ovary trehalase of the silkworm, *Bombyx mori* and its transcriptional activation by diapause hormone. *Biochem. Biophys. Acta.*, 1218: 366-374.

Subramanian, K. and Murthy, N.C.V. (1987) Effect of rest on the efficiency of male moth in relation to fecundity and hatching behaviour of silkworm eggs. *Sericologia*, 27: 677-680.

Sugai, E. and Hanaoka, A. (1972) Sterilization of the male silkworm, *Bombyx mori* (L.) by the high temperature environment. *J. Seric. Sci. Japan*, 41: 51-56.

Sugai, E. and Kiguchi, K. (1967) Male sterility of the silkworm, *Bombyx mori* L. induced by high temperature during pupal period. *J. Seric. Sci. Japan*, 36: 491-496.

Sugai, E. and Takahashi, T. (1981) High temperature environment at the spinning stage and sterilization in the males of the silkworm, *Bombyx mori* L. *J. Seric. Sci. Japan*, 50: 65-69.

Suzuki, K.; Fujita, M. and Miya, K. (1983) Changes in super cooling of silkworm eggs. *J. Seric. Sci. Japan*, 52: 185-190.

Suzuki, K.; Hosaka, M. and Miya, K. (1984) The amino acids pool of *Bombyx mori* eggs during diapause. *Insect Biochem.*, 14: 557-561.

Suzuki, K. and Miya, K. (1975) Studies on the carbohydrate metabolism in diapause eggs of the silkworm, *Bombyx mori* with special reference to phosphofructokinase activity. *J. Seric. Sci. Japan*, 44: 88-97.

Suzuki, N.; Okuda, T. and Shinbo, H. (1996) Sperm precedence and sperm movement under different copulation intervals in the silkworm, *Bombyx mori* L. *J. Insect Physiol.*, 42: 199-204.

Suzuki, K.; Tsuda, S. and Miya, K. (1978) Occurrence of non-diapause eggs by injection of odenosine-5'-monophosphate into *Bombyx mori* pupae. *Appl. Ent. Zool.*, 13: 48-50.

Takami, T. (1946) Experimental studies on the embryo formation in *Bombyx mori*. V. Presumptive mesodermal and neural regions of the eggs. *Seibutu*, 1: 208-211.

Takami, T. (1969) '*A General Textbook of the Silkworm Eggs*'. Zenkoku Sanshu Kyokai, Tokyo, Japan.

Takami, T. (1972) *In-vitro*: Development of insect embryos; in *Invertebrate Tissue Culture*. Vago, C. (ed.), Vol. II., pp. 137–159, Academic Press, New York.

Takami, T. and Kitazawa, T. (1960) External observation of embryonic development in the silkworm. *Sanshi, Shikenjo, Hokoku, Sericult. Exp. Stn. Tech. Bull.*, 75:1-31.

Takami, T.; Totani K.; Sugiyama H.; Kitazawa T. and Kanada T. (1966) Embryonic growth of the silkworm, *Bombyx mori* (L) at the early post-diapause stage with special reference to the connection between the growth and nuclear division. *Bull. Sericult. Exp. Stn.*, 20: 57–69.

Takeda, S. (1977a) Stage dependency of diapause egg production by corpora cardiaca and corpora allata complex of the silkworm, *Bombyx mori* (Lepidoptera: Bombycidae). *Appl. Entomol. Zool.*, 12: 80-82.

Takeda, S. (1977b) Induction of egg diapause in *Bombyx mori* by some cephalothoracic organs of the *Periplanata americana*. *J. Insect Physiol.*, 23: 813-816.

Takeda, S. and Hasegawa, K. (1975) Alteration of egg diapause in *Bombyx mori* by quabain injection into egg diapause producers. *J. Insect Physiol.*, 21: 1995-2003.

Takeda, S. and Hasegawa, K. (1976) Further studies on the alteration of egg diapause in *Bombyx mori* by quabain-diapause egg production in non-diapause egg producers. J. *Seric. Sci. Japan*, 45: 337-344.

Takesue, S.; Keino, H. and Onitake, K. (1980) Blastoderm formation in the silkworm eggs (*Bombyx mori* L). *J. Embryol. Exp. Morphol.*, 60: 117-124.

Takizawa, H.; Vivier, E. and Pettprez, A. (1975) Recherches cytochiniques sur la microsporidie *Nosema bombycis* ancours de son development chez lever a soil (*Bombyx mori*). *J. Protozool.*, 22, 359-368.

Talukdar, J.N. (1980) Prevalence of transovarian infection of microsporidian parasite infecting muga silkworm, *Antheraea assamensis*. *J. Invertebr. Pathol.*, 36: 273-275.

Tanaka (1964) *Sericology*. Published by Central Silk Board, Bangalore, India.

Tatsuke, K. (1971) Pebrine disease of silkworm – A technical report. Overseas Technical Cooperation Agency, Tokyo-Japan, pp. 1-20.

Tazima, Y. (1951) Separation of male and female silkworm in egg stage becomes possible. *Silk Digest*, Tokyo, 61.

Tazima, Y. (1964) *The Genetics of the Silkworm*. Logos Press, London.

Tazima, Y. (1978) '*Silkworm Egg*'. Published by Central Silk Board, Bangalore, India, pp. 1–49.

Turnock, W.J. and Fields, P.G. (2005) Winter climates and cold-hardiness in terrestrial insects. *European J. Entomol.*, 102: 561-576.

Undeen, A.H. and Cockburn, A.F. (1989) The extraction of DNA from microsporidian spores. *J. Invertebr. Pathol.*, 54: 132-133.

Upadhyay, V.B. and Mishra, A.B. (1974) Influence of temperature on the passage of food through gut of multivoltine *Bombyx mori* larvae. *Indian J. Seric.*, 33: 183-185.

Vavra, J. and Maddox, J.V. (1976) Methods in microsporidiology; in *Comparative Pathobiology*. Bulla, L.A. and Cheng, T.C. (eds.), Vol. I., pp. 107-121, Plenum Publishing Corporation, New York.

Veber, J. (1958) A comparative histopathology of the microsporidians, *Nosema bombycis* on different hosts; in *Transactions of the First International Congress of Insect Pathology and Biological Control*, pp. 301-314, Praha.

Venkatareddy, S.; Singh, B.D.; Baig, M.; Sengupta, K.; Giridhar, R. and Singhal, B. K. (1990) Efficacy of asiphor as a disinfectant against incidence of diseases of silkworm, *Bombyx mori*. *Indian J. Seric.*, 29: 147-148.

Verma, M. and Chauhan, T.P.S. (1996) Effect of long-term preservation of eggs on hatching and rearing performance of some polyvoltine silkworm races, *Bombyx mori* L. in Northern India. *Sericologia*, 36: 249-253.

Vossbrinck, C.B.; Maddox, J.V.; Friedman, S.; Debrunner-Vossbrink, B.A. and Woese, C.R. (1987) Ribosomal RNA sequence suggests microsporidians are extremely ancient eukaryotes. *Nature*, 326: 411-414.

Wechsung, E. and Houvenaghel, A. (1976) A possible role of prostaglandins in the regulation of ovum transport and oviposition in the domestic hen. *Prostaglandins*, 12: 559-608.

Weiser, J. (1969) Immunity of insects to protozoan; in *Insects Immunity to Parasitic Animals*. Jackson, C.J.; Herman, R. and Singer, I. (eds.), pp. 129-147, Appleton Century Crofts, New York.

Weiser, J. (1977) Contribution to the classification of microsporidia. *Vestn Cesk Spol Zool.*, 41: 308-321.

Wigglesworth, V.B. (1972) *Insect Hormones*. Oliver and Boyd, Edinburgh.

Wilde-De, J. and Loof-De, A. (1973) Reproduction; in '*The Physiology of Insects*'. Rockstein, M. (ed.), pp. 11–95, Academic press, New York.

Wright, S. (1921) Systems of mating. V. General considerations. *Genetics*, 6: 167-168.

Wu, C.; Yamashita, T.; Taira, H. and Suzuki, K. (1996) Free nucleotides related to the induction of diapause in the silkworm, *Bombyx mori* L. *J Insect Physiol.*, 42: 441-447.

Yaginuma, T. and Yamashita, O. (1977) Changes in glycogen, sorbitol and glycerol content during diapause of the silkworm eggs. *J. Seric. Sci. Japan*, 46: 5–10.

Yaginuma, T. and Yamashita, O. (1978) Polyol metabolism related to diapause in *Bombyx mori* eggs, different behavior of sorbitol from glycerol during diapause and post-diapause. *J. Insect Physiol.*, 24: 347-354.

Yaginuma, T. and Yamashita, O. (1979) NAD-dependent sorbitol dehydrogenase activity in relation to the termination of diapause in eggs of *Bombyx mori. Insect Biochem.*, 9: 547–553.

Yaginuma, T.; Kobayashi, M. and Yamashita, O. (1990a) Distinct effects of different low temperatures on the induction of NAD-sorbitol dehydrogenase activity in diapause eggs of the silkworm, *Bombyx mori. J. Comp. Physiol.*, 8: 160-277.

Yaginuma T.; Kobayashi, M. and Yamashita, O. (1990b) Effects of low temperatures on NAD-sorbitol dehydrogenase activity and morphogenesis in non-diapause eggs of the silkworm, *Bombyx mori. Comp. Biochem. Physiol.*, 97 (B): 495–506.

Yamashita, O. (1984) Control of embryogenesis and diapause in the silkworm, *Bombyx mori*: Roles of diapause hormones and egg specific proteins. *Adv. Invert. Rep.*, 3: 251-258.

Yamashita, O. and Hasegawa, K. (1985) Embryonic Diapause; in *Comprehensive Insect Physiology, Biochemistry and Pharmacology.* Kerkut, G.A. and Gilbert, L.I. (eds.), Vol. I., pp. 407-434, Pergamon Press, Oxford.

Yamashita, O. and Irie, K. (1980) Larval hatching from vitellogenin -deficient eggs developed in male hosts of the silkworm. *Nature,* 283: 385 -386.

Yamashita, O.; Suzuki, K. and Hasegawa, K. (1975) Glycogen phosphorylase activity in relation to diapause initiation in *Bombyx mori* eggs. *Insect Biochem.,* 5: 707–718.

Yamashita, O. and Yaginuma, T. (1991) Silkworm eggs at low temperatures: Implications for sericulture; in *'Insects at Low Temperature'*. Lee, R.E. and Denlinger, D.L. (eds.), pp. 424–445, Chapman and Hall, New York.

Yamashita, O.; Yaginuma, T. and Hasegawa, K. (1981) Hormonal and metabolic control of egg diapause of the silkworm, *Bombyx mori* (Lepidoptera: Bombycidae). *Ent. Gen.,* 7: 195–211.

Yamashita, O.; Yaginuma, T.; Kobayashi, M. and Furasawa, T. (1988) Metabolic shift related with embryonic diapause of *Bombyx mori*: Temperature directed sorbitol metabolism; in *Endocrinological Frontiers of Physiological Insect Ecology*. Sehnal, F.; Zabza, A. and Denlinger, D.L. (eds.), pp. 263-275, Wroclaw Tech. Univ. Press, Wroclaw.

Yamayoka, K. and Hirao, T. (1973) Releasing signals of oviposition behavior in *Bombyx mori. J. Insect Physiol.,* 19: 2215-2223.

Yamayoka, K. and Hirao, T. (1981) Mechanisms of ovipositional behaviour in *Bombyx mori*: time gating and accumulation of the internal factor. *Intl. J. Invertebr. Reprod.,* 4: 169-180.

Yamayoka, K.; Hashino, M. and Hirao, T. (1971) Role of sensory hairs on the anal papillae in oviposition behaviour of *Bombyx mori. J. Insect Physiol.,* 17: 871-879.

Yeole, R.G.; Sarnaik, D.N. and Satpute, U.S. (1995) Effect of mating duration on fecundity and egg viability of silkworm, *Bombyx mori* L. *J. Seric.* (India), 3: 80-82.

Yi, S.X. and Lee, R.E. Jr. (2005) Changes in gut and Malpighian tubule transport during seasonal acclimatization and freezing in the gall fly, *Eurosta solidaginis*. *J. Experimental Biology*, 208: 1895-1904

Yokoyama, T. (1963) Sericulture. *Ann. Rev. Ent.*, 8: 287-306.

Yokoyama, T. (1973) The history of Sericulture Science in relation to industry; in *History of Entomology*. Smith, R.E.; Miller, *I.E.* and Smith, C.N. (eds.), pp. 267-284

Yokoyama, T. and Sugai, E. (1987) Effect of hot water treatment on early development of eggs of the silkworm. *J. Seric. Sci. Japan*, 56: 441-442.

Yokoyama, T.; Sugai, E. and Oshiki, T. (1987) Effect of acid treatment on early development of eggs of the silkworm. *J. Seric. Sci. Japan*, 56: 369-373.

Yu, S.J.; Chen, J.S. and Hsieh, F. K. (1993) Studies on preservation of eggs of the multivoltine silkworm, *Bombyx mori* L. *Zhongva Kunchong.*, 13: 141-149.

Yu, S.J.; Hsieh, F.K.; Hou, F.R. and Chu, H.T. (1991) Effect of acid treatment on the hatchability of *Bombyx mori* eggs at different embryonic stages. *Chinese J. Entomol.*, 11: 13-18.

Zachariassen, K.E.; Kristiansen, E.; Pedersen, S.A. and Hammel, H.T. (2004) Ice nucleation in solutions and freeze-avoiding insects–homogeneous or heterogeneous? *Cryobiology*, 48: 309-321.

Zhu, J.; Indrasith, L.S. and Yamashita, O. (1986) Characterization of vitellin, egg-specific protein and 30-kDa proteins from *Bombyx mori* eggs and fats during oogenesis and embryogenesis. *Biochem. Biophy. Acta*, 882: 427–436.

Ziegler, R.; Ashida, M.; Fallen, A.M.; Wimer, L.T.; Wyatt, S.S. and Wyatt, G.R. (1979) Regulation of glycogen phosphorylase in fat body of Cecropia silk moth pupae. *J. Comp. Physiol.*, 149: 321-332.

Index